Matthias Schrader

TRANS FORMATIO NAL PRO DUCTS

*The code behind digital products that are shaping our lives
and revolutionizing our economies*

NEXT FACTORY OTTENSEN — NFO/00/TP

ABOUT THE AUTHOR

Matthias Schrader is one of Europe's digital pioneers. He founded digital marketing and advertising agency SinnerSchrader in the mid-1990s and began to develop e-commerce solutions for startups like Intershop, Ricardo and buecher.de, enabling them to go public earlier than expected.

In 1999, SinnerSchrader issued a public offering of its own and was one of only a handful of fledgling companies that survived the so-called "new economy" and actually emerged strengthened by the experience. In 2006, Schrader founded the NEXT Conference which has become the leading symposium for → digital transformation in Europe.

Today, the author, with his team of more than 500 consultants, designers and software engineers, continues to assist many large DAX index companies to develop cutting-edge digital products. In February 2017, the worldwide management and technology consultancy Accenture announced it was taking a controlling interest in SinnerSchrader for a nine-figure sum.

ACKNOWLEDGEMENTS

A book like this never just happens. In fact, I was lucky to be able to stand on the shoulders of many more-erudite authors, many of whom I have listed in the appendix. Martin Gassner (product design) and Holger Blank (product engineering) have contributed entire passages.

Klaus-Peter Frahm, Michael Schieben and Wolfgang Wopperer-Beholz introduced me to the Product Field method, a comprehensive model and tested toolbox for → Product Thinking, in 2016 and I am grateful to them for providing the corresponding chapter for this book.

Martin Recke spent weeks producing the transcripts of the series of interviews we conducted in the late summer of 2016 – before we decided to start over again. Without his perseverance, this book would never have been born. But, above all, I want to thank the terrific team of dedicated individuals who not only kept the shop running while I was busy writing, but also inspired and supported me through many, many conversations. Special thanks to Axel Averdung.

Last, but not least I have to thank Tim Cole for the english translation and Eric Doyle for his careful editing. Their work greatly improved this book and helped to make it available to an international audience.

Thank you all!

Hamburg, September 2017

Table of Contents

User Manual

This book comprises three distinct parts. The first describes the rise of the → Casual Economy over the course of the past three decades. How did Google, Apple, Facebook and Amazon manage to become so dominant and what methodologies did they follow to create their Transformational Products? If you're in a hurry and want to use this book quickly, put this part aside and read it at your leisure on your next vacation or business trip.

The second part of the book describes the specific characteristics – in short, the code – behind Transformational Products and discusses what it takes to make them succeed. The third and final section is intended as a playbook enabling you to develop successful products of your own. This is also where we bridge the gap between product development and the digital transformation of entire companies.

Finally, be aware that this book is written in Geek; a language ripe with terms and expressions that have a digital context. In addition, it is our intent to introduce a model to be used by mixed teams – managers, designers and engineers require a common terminology for their methodic approach. Don't worry, we have assembled all the important terms in a glossary at the end of the book, and when a new term is introduced in the text you will find it → underlined.

PROLOGUE

From zero to 36 billion

"We shall meet in the place where there is no darkness."

– George Orwell, 1984 (1948)

It is the Orwell Year, 1984. On my desk stands a Commodore 64 with an acoustic coupler linked to a telephone via simple handset cups purchased at the local hardware store. At the other end of the copper twisted-pair line another home computer is connected, and the two of them "shake hands" virtually through an exchange of pips and whistles, modulating and receiving data sent from one PC to the other. I'm online for the very first time! Looking back over 30 years, I could never have imagined that this was the beginning of the Digital Age. Back then, in the mid-1980s, there were only a few thousand geeks of my generation in the whole world able to mesh their computers for a few minutes per day. Today, we're always online, anytime, anywhere. Daily, the internet is entering deeper into our lives and doing new chores for us. It has become the greatest convenience machine ever invented.

But the net is not only reshaping our lives, it is also transforming the way businesses do business. How did this come to pass, and why are the effects on companies and products so breathtaking?

This web of connectivity didn't happen overnight. In fact, it took decades to come about, and it proceeded in a series of waves and tiny steps. The first occurred in 1977 when Steve Jobs and Steve Wozniak developed and introduced their Apple II, the first pre-configured personal computer. Four years later, IBM followed suit with the very first PC to bear that name. Bill Gates of Microsoft provided the operating system for the IBM machine and he made sure he retained the licensing rights to what he called MS-DOS. Nobody at IBM, apparently, realized just how important the PC would become. In no time at all developers at other companies were turning out all kinds of word processors and spreadsheets.

The PC changed the office. It became everybody's little helper. Two decades later in 2001, shipments of new PCs reached 130 million units in a single year.

The second wave was even higher and faster: The world wide web needed only 15 years from its beginnings in the mid-1990s to reach de-facto saturation. In 2007, when Apple CEO Steve Jobs presented the first iPhone to the public, the number of internet users worldwide passed one billion. Eventually, the mobile wave surpassed even those numbers, with more than 36 billion sim cards produced in 2017 alone. The smartphone has become the world's remote control around which many of our everyday functions are organized.

Welcome to the Casual Economy.

Kodak Moments

"As one industry after another looks at itself in the mirror and asks about its future in a digital world, that future is driven almost 100 percent by the ability of that company's product or services to be rendered in digital form."

– Nicholas Negroponte, Being Digital (1995)

That binary digits (bits) have certain advantages over physical materials (atoms) is well-established. Nicholas Negroponte coined the phrase "bits vs atoms" long ago in his bestseller *Being Digital*. The context was clear even then. Bits are orders of magnitude cheaper to process and distribute than atoms. Thus, digitizing processes, business models and products yield returns even if they otherwise remain fundamentally unchanged.

But that isn't how it turns out in most cases because connectivity alters the underlying business logic and change is inevitable. Reducing distribution and transaction costs essentially to zero opens the way to completely new business models. In addition, the marginal costs, such as the additional sum incurred in the production of one more unit of a product or service, also falls to nothing, so digital products are not subject to the laws of scarcity. Once produced these goods can be reproduced any number of times at essentially no cost at all. Seen that way, the internet is actually one great big copying machine.

Anyone trying to digitize a business model based on scarcity is up against a fundamental problem. This can only work with products that are not easy to substitute; otherwise the attempt will fail. Today, it's users and attention spans that are scarce, not the digital products themselves. That is the reason for the power shift within the digital economy to the customer and away from the vendor.

All this first became apparent in the mid-1990s when e-commerce began to revolutionize retail markets thanks to unbeatable distribution and transaction costs. However, the products themselves remained basically unchanged. On the other hand, the internet is now transforming the products by placing itself in the center of the product experience, as network effects increasingly become part of the product's perceived value. More and more products are becoming apps for smartphones. Those that offer the best → user experience (UX) catch on quickly, reach more users than their competitors and, over time, squeeze them out of the market. The user experience offered by Uber not only beats old-fashioned taxi rides, it can also potentially make owning a car undesirable. The atoms of the car become less important for the product "mobility" than bits represented by the Uber network. Digital products enrich the user through the experience of using them, while traditional products gradually fade into legacy, losing their interface and their customer access.

Digitization demonstrates the validity of a new branch of economics called → service-dominant logic, or S-D logic, which describes all economic activity as a service-for-service exchange in which the activities people want done for them represent the source of value and thus the purpose of exchange, not the goods, which are only occasionally used in the transmission of the service. In other words, the value for the consumer does not depend on the product itself but on its → utility, which is something created by the customer. The service is the product, and it can be reproduced much more cheaply and efficiently than atoms. Digital services can also be improved and upgraded much faster than physical products.

Essentially, a digital service is just software, and software innovation cycles are naturally much shorter than those for hardware. Amazon updates its software during peak times more than 1,000 times an hour. Software updates are not only faster than exchanging hardware but also add new value to the hardware. In every automobile it produces, Tesla already installs the hardware which will one day enable their cars to drive themselves, even though the necessary software isn't available yet. They plan to offer autonomy later as an update for which, of course, they will charge extra.

→ Functions-on-demand are the strongest indication we have that the paradigm shift from atoms to bits is already well under way. Without software, hardware is progressively declining to offer zero value. Another factor that is contributing to this obsolescence is virtualization. In this case, hardware is completely replaced by software which makes more efficient use of physics. The hardware itself is increasingly being moved to the cloud.

In his bestseller *The Innovator's Dilemma*, published in 1997, Clayton Christensen was first to point out this obvious contradiction, namely that it is smarter for companies to concentrate on their most profitable customers and products. It is better, he argued, to ignore disruptive technologies if they do nothing to better satisfy the needs of their key customers, or that do not fit their current business model. New technologies, it turns out, are often only marginally superior to existing ones, and often, they may prove inferior. That's why some new technologies only enter the bottom of the market or exist in expensive niches.

It's hard to predict how this will turn out. In the digital world, a multitude of parameters – processing power, storage capacity, bandwidth – play a role, with each evolving exponentially. Each generation is twice as powerful as its predecessor, and since each new technology basically starts at zero, the hardest step is the very first one – from *Zero to One*, as Peter Thiel, one of the founders of PayPal, wrote in the book bearing that title. Systems that are subject to exponentiality grow slowly at first. It takes 10 doublings to reach a thousand, but then things start happening extremely quickly. Another 10 doublings, and you reach a million. Another 10, and we are suddenly talking billions. However, that is just theory. In the real world, a bunch of factors combine to slow development down: saturation, for instance, and disruptive competition. Technological progress alone does not make robust enterprises. Andy Grove, the co-founder who turned Intel into the world's most successful manufacturer of computer chips, was right when he titled his management bestseller *Only the Paranoid Survive.*

As long as a product and its business model work, the pressure to transform is fairly limited. Big companies are especially good at systematically and incrementally improving products and business models. For decades, Kodak was a perfect example. It always spent loads of money on R&D, ran huge laboratories and invested heavily in innovation, but in the end, it was swamped by digitization which made its chemical film business obsolete. The old ways simply worked for too long, despite the fact that Kodak succeeded year by year in making its product better and better.

Kodak's rejection of digital cameras was not necessarily based on arrogance alone. Neither the company nor the people working for it were stupid – quite the opposite because it's usually the smart people who seek to find out what will work and what won't. But when Kodak asked its customers what they wanted, they got answers like: "We want a film that produces even brighter colors, can be processed ever faster, and is even more impervious to the kinds of changes in lighting that frequently occur in photography."

Kodak was very good at improving these properties in its products over time and customers, after all, were unable to imagine that taking pictures could be done any other way – with the help of digital chips, for instance. And, at least in the beginning, digital pictures were dramatically inferior to chemical photographs which were the result of processes refined over decades. Old products are generally superior to disruptive newcomers pushing in from the sidelines, at least when they have reached their zenith of development.

The first commercially available digital cameras had a much lower resolution than a single-lens reflex camera using traditional film – just as electric cars are inferior to internal combustion models in terms of range and price. YouTube videos used to be much grainier than television: they jerked, the screen was a lot smaller, and the picture quality was dismal. Over time, streaming video eventually gained the upper hand over broadcast TV. Digital products have a big advantage: The infrastructure on which they are based can grow exponentially, as we will see in the next chapters. This allows them to profit from the so-called "network effect", which opens up a whole new dimension of benefits.

Gordon Moore, another co-founder of Intel, forecast way back in the 1960s that the density of the transistors in integrated circuits would double every 12 months or so. Known today as Moore's Law, this principle still governs progress in the digital world, even if for practical purposes it has slowed down slightly, now doubling only every 18 months on average. This means that while a standard microprocessor will become twice as powerful, the price will remain the same.

The exponential growth of computer power also leads to a dramatic drop in price for processing, storage, sensors, and bandwidth. What we call the Digital Age is really the Age of Connectivity. Thanks to falling prices networks could be extended from big mainframe computers to more personal ones and then to smartphones and soon will incorporate just about everything through the ubiquitous → Internet of Things. Linking a gadget up to the internet costs just a few cents. This mixture of ever more powerful, fully networked devices and a global cloud computing infrastructure are leading to the explosive growth of new services and products.

Long before digitization, enterprises struggled (and failed) to cannibalize their legacy business because the result was usually not as good, and hence less profitable, than the old one. It is therefore considered an established fact that disruptive innovation needs to come from outside because large companies are blind to transformation – and in fact probably need to be so. But is this really true? Do companies have to sit still and wait for some outsider to turn up and crash their business model? Why can't they create their own disruptive innovation?

Upheavals

"An iPod, a phone, and an internet communicator. An iPod, a phone... Are you getting it?"

– Steve Jobs, iPhone Launch (2007)

However, startups struggle, too, and many don't survive, despite being completely digital, having digital natives at the helm and being steeped in the digital culture. If all companies had the same success rate as startups, our economy would have tanked years ago – statistically, at least.

If it's not the culture and the methods, what is it the difference that makes the difference? It's the product, stupid! Products with the potential to transform customer behaviors, markets and enterprises. It takes products that create value for customers in today's digital ecosystem to ensure the future of your company.

The time a company spends on the S&P 500, Standard & Poor's index of the 500 largest companies on the stock exchange, fell to 18 years in 2012 – in 1980 it was 25 years, and in 1958 it was 65 years. If this trend continues, three quarters of today's S&P 500 companies will have completed the full circle of rise and fall on this major stock market index and will be gone by 2027.

This shows graphically how hard it has become to stay relevant over a prolonged period. If corporations fail, it's because their products have become irrelevant. Nokia and Blackberry, for instance, were doomed the day Steve Jobs took the stage to introduce the iPhone in 2007.

In order to improve their chance of survival established companies need to create a pipeline of Transformational Products. Unlike startups, large companies can't bet on a single horse. They need a mix of in-house development, partnerships and acquisitions. This resembles the way successful venture capitalists build up a portfolio of startups.

In transforming itself into Alphabet, Google placed a number of substantial bets because its executives realized their existing product portfolio would one day become obsolete. Larry Page and Sergey Brin are putting their money on Alphabet's ability to create new blockbuster products, and it's Alphabet's job to develop them and stuff them in the pipeline.

Successful products transform the way consumers use them. They are habit-forming products that supplant existing habits, as Nir Eyal explained in his book *Hooked* (2014). Larry Page routinely subjects every new product to the toothbrush test: Is this something I would use at least once or twice a day, and does it improve the quality of my life? It's all about relevance to everyday users and about changing their behavior.

In the digital age, the value chain is turned on its head. The greatest value is added at the intersection with the user – and that's why control of the → user interface is so crucial. As customer and consumer behavior changes, so do markets, and eventually, if successful, the company. Corporate change happens at the end of the process, not at the beginning. Digital transformation affects the user first, then the market and, finally, the company.

Developing Transformational Products is no trivial task. The key is finding new benefits for customers, finding the right shape for the product and designing the right business model. It's hard to plan this ahead of time, and the path to success is often long and winding. It usually involves lots of trial and error, and endless testing. That's a lot of effort with costs both in time and money, but even so can sometimes end in failure.

That's the reason many corporations prefer to concentrate on easier stuff like marketing, sales, and procurement, control processes that require good planning. In short, they focus on incremental improvements and

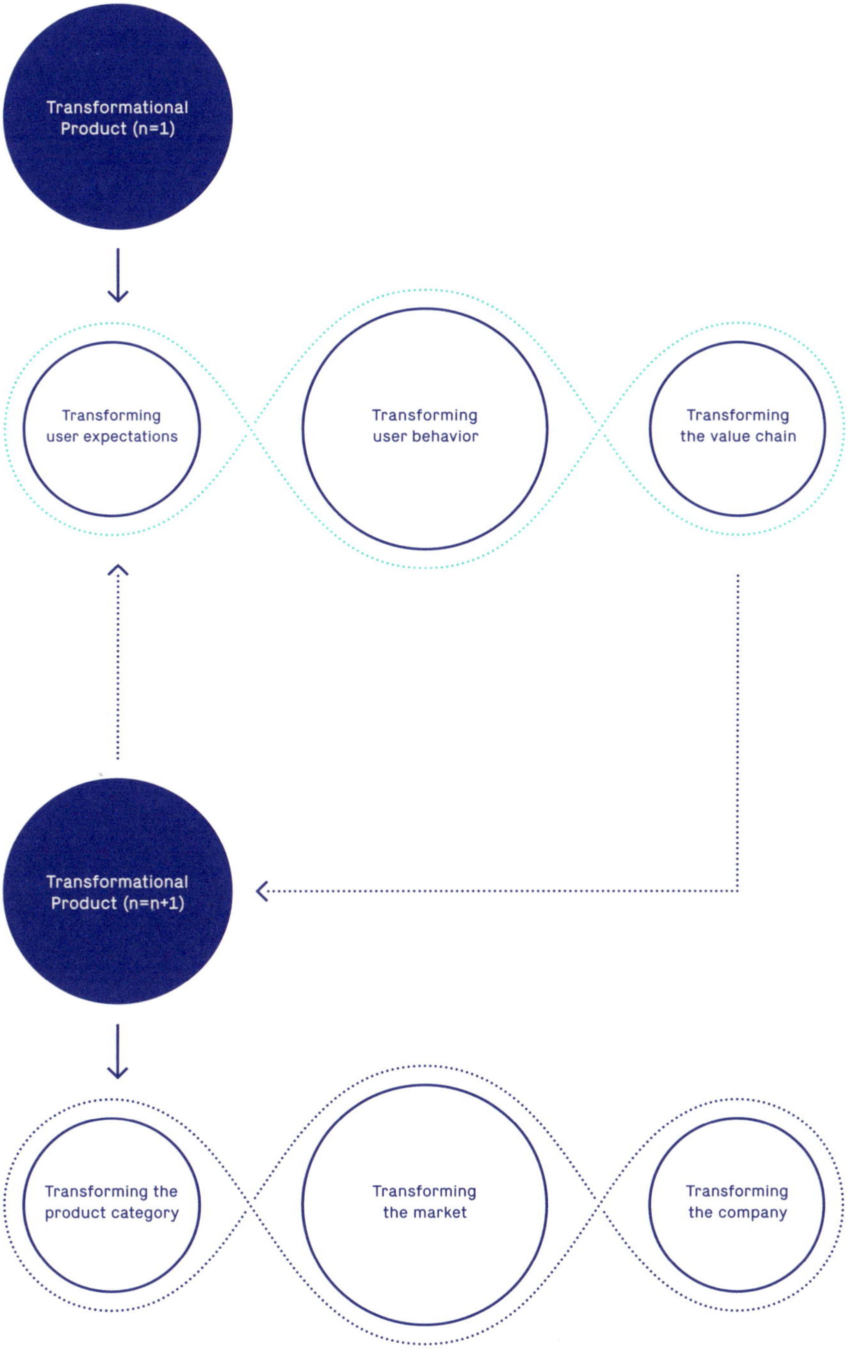

Fig. 1: Digital transformation

The diagram shows:

- Transformational Product (n=1)
- Transforming user expectations
- Transforming user behavior
- Transforming the value chain
- Transformational Product (n=n+1)
- Transforming the product category
- Transforming the market
- Transforming the company

risk-avoidance. As Peter Drucker, the pioneer of modern management theory, said: successful companies are good at doing things the right way, but in times of transformation and upheaval, it is more important to do the right things.

This is what this book is about.

The rise of the personal computer

MOORE'S LAW REVISITED

For more than 50 years → Moore's Law has been the yardstick against which digital transformation is measured. The passing of seven Moore cycles – roughly 10 years – means a processor that was once worth a thousand dollars would now cost less than a bag of coffee. Wait another seven Moore cycles and it will only cost a dime. Price isn't the only thing that shrinks, the size of the hardware keeps getting smaller, too. A Raspberry Pi, for instance, is a fully functional PC but it only has the footprint of a credit card. In fact, the computing power of a PC you could purchase in 1999 is available today for a few cents and is ready to be built into all kinds of everyday devices and appliances.

However, you can't simply map the exponential growth of microchips onto every business model or use it to predict how a certain product will perform in the market. In most cases progress generally follows a sigmoidal, or S-shaped, curve. Along with the exponential growth curve which keeps getting steeper and steeper, say the number of transistors you can cram into a given microchip, there are limiting factors working in the opposite direction.

When introducing new products that factor is total market capacity. Even for a product that is theoretically attractive to everyone on the planet there eventually comes a point when it hits a saturation limit.

S-curve growth starts off slowly. PC development initially kicked off with the Altair 8800 (1975) and the Apple II (1977) and initial sales followed a very shallow curve. This was also true when the first wave of home computers appeared, fueled by vendors such as Commodore, Texas Instruments and Sinclair. In 1981, IBM launched its first PC, bringing computers out of the hobby cellar into the office. This development reached its pinnacle in the mid-1990s, so the S-curve took about 20 years to climax.

The introduction of the world wide web followed a similar path between 1995 and 2010. Only this time, the curve took just 15 years to achieve maturity, reaching two billion users worldwide in June 2010. The mobility cycle, which we are in the middle of right now, will progress even faster. Most experts predict the wave will top off around 2020 with virtually every person on the planet above the age of 13 owning a smartphone.

Comparing these curves proves that things are speeding up. Each technological generation profits from Moore's Law leading to improved performance and lower costs. Originally, an IBM PC cost the equivalent of two month's pay for an average worker but, around the turn of the century, consumers only needed to plonk down a third of a month's salary to buy a fully-loaded notebook. A smartphone, especially if it is cross-funded by a telephone company, goes for a single Euro. An Apple Watch in 2017 boasts the same computer power as a Cray-2 supercomputer back in the mid-1980s which was then the fastest – and most expensive – computer in the world.

As for smartphones, we seem to be approaching the peak point on the curve again. The next S-curve is almost in sight.

THE FOUR CYCLES OF PERSONAL COMPUTING

Cycle #1: Office and hobby (1975-1995)

When Jeff Bezos founded Amazon in 1994, the internet was just a niche computer application because you had to install a modem to access it. In addition, the PC craze of the second half of the 1970s was gradually losing steam. Apple, Microsoft and Intel had successfully established computers in offices around the world, as well as in hobby shacks and dens, but these markets were starting to saturate.

Cycle #2: Web and e-commerce (1995-2010)

The second wave began in the mid-1990s with the success of the web, and e-commerce became its first killer application. From 1996 to 2001 online retailers had their initial heyday, but all the vendors did, basically, was to produce a digital version of a well-known business model, namely distance selling. Big mail-order companies like Sears Roebuck had been doing this successfully since the end of World War II. Electronic catalogs simply replaced their printed predecessors.

What really changed, thanks to the web, was marketing. Online, the "four Ps" (product, price, promotion, and place) underwent various degrees of transformation. Initially, digitization only effected the last P – place; specifically, the site of the transaction along with the corresponding distribution channels. This has since spread to impact virtually every branch of retail, starting with the very first e-commerce vendors. The rest of the economy remained basically unchanged; the "digital shockwave" passed them by initially. Mail-order market share remained stable at about 15 percent in most developed countries, and e-commerce took more than 10 years to draw level with and then surpass traditional distance sellers – wreaking dramatic change in virtually every retail sector.

The first to feel the pressure were traditional mail-order giants like Sears in the US and Otto, Quelle and Neckermann in Europe. At first, these vendors were able to reap significant rationalization effects from the web.

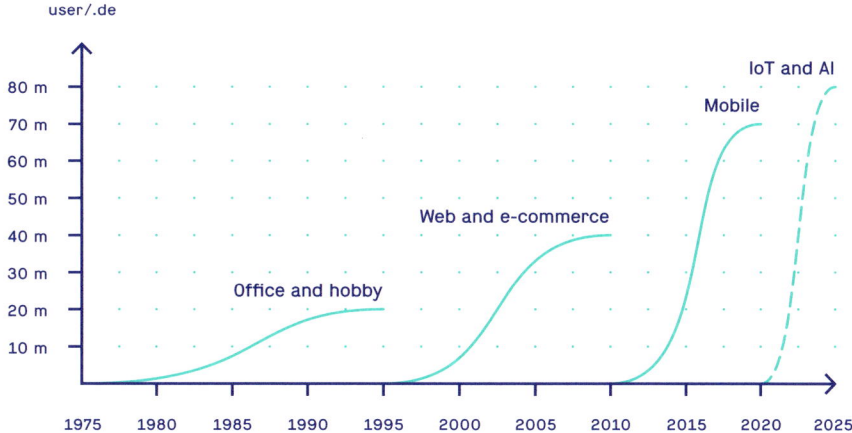

Fig. 2: The four cycles of personal computing

For them, e-commerce was just a continuation and a strengthening of their tried and proven business model. There seemed to be no sign of any real disruption – but it had already set in, spearheaded by a tiny intrepid startup in Seattle called Amazon. Jeff Bezos had founded Amazon to fulfill his ambitious vision to become the biggest retailer in the world. He could not have cared less what he sold and for him it was only important to get to first base. Books seemed like a good place to start for any number of reasons. Books were easy to handle logistically and were read by millions of people.

The best thing was that the book market operated with intermediary suppliers who kept millions of titles in stock and could deliver them to any bookseller in the land, including Amazon, within days. Bookshops sold books to consumers even though they didn't actually have them in stock, relying on their suppliers to deliver them within a day or two. The customer would visit a bookstore, already knowing what they wanted, and if it wasn't on their shelves, the store would procure it by referring to the suppliers' catalogs listing all the books available for sale. These catalogs were initially paper books, but but from the early 1990s onwards digital editions became increasingly available.

At the time, Amazon relied on the listings of Ingram Books, one of its inter-mediaries in the US. By standing on the shoulders of its supplier, Amazon could take advantage of Ingram's catalog and logistics, even though it was still essentially a garage company. Soon Amazon was able to claim, even in court, to be the largest bookseller in the world. And that was just for starters.

Catalogs and logistics were all you needed for e-commerce in the 1990s, but only a few retail sectors were involved and the process was achingly slow. But that didn't matter because there weren't that many internet users anyway, so online demand for goods remained limited. It was only when more people connected with the internet and got used to buying online that e-commerce really took off. For others, it didn't seem such a big deal at first. Sure, bank-ing, insurance, automotive and telecommunications were watching what was going on, but in general they viewed the whole digitization thing as a niche topic.

That all changed overnight when the internet boom set in just before the turn of the century. Venture capitalist investors turned the first IPOs of Netscape and Yahoo into the financial equivalent of Oscar Night and poured unprecedented millions into tiny startups whose stock prices broke every record. While investors were falling over each other to put up money for these new stock market darlings, consumers proved slower to change their habits and the bubble finally burst in 2001. It turned out that too many dotcoms had spent their new-found riches on marketing hype and not enough on creating → viable products, pushing half-baked products into the market even though customers weren't ready for them because sales were still on the wrong end of the S-curve. Managers in traditional businesses like retail were relieved, believing all the digital ballyhoo was just so much baloney and that the internet would always remain a sideshow.

Over time, people caught up and started using the internet more inten-sively. The net had become the more friendly web and had attracted a wider and wider audience. Everybody seemed to want a piece of the action. Compa-nies like eBay helped to empty our attics, Amazon filled our mailboxes with books and CDs, and Travelocity found us the best deals on flights to islands in the sun. Autobytel became the most popular place to shop for your next car. Suddenly email addresses from Yahoo! or Hotmail started cropping up not just on business cards but also on personal ones.

The dotcom bubble may have burst for the investors but for consumers the fun had just started. Internet usage reached and surpassed its → tipping point as more and more people bought PCs so they, too, could access the wonderful world of "cyberspace". Web browsers were no longer considered a PC gimmick and had become the quintessential killer app. The PC itself cast off the cables that had bound it to the desktop to become notebooks and escape from the study to invade our living rooms, kitchens and bedrooms. Thanks to WiFi these first mobile computers were "always online". Digital subscriber line (DSL) and cable providers rolled out flat rate deals; no longer was internet access measured in bytes and minutes as in the early days of online service providers like CompuServe and AOL.

The web became part of everyday life as people increasingly spent time in front of a completely different screen to the familiar television and, like the TV set, every household had to have its own internet connection as more and more everyday chores started to become digitized. Banking turned into online banking and your friendly neighborhood insurance salesperson was replaced by web-based comparison engines which could give a quote on the best policy deals within seconds. Purchasing decisions became Google searches – more and more of which led directly to online shops.

Cycle #3: Mobile (2010-2020)

Exactly 30 years after the advent of the Apple II computer, Steve Jobs kicked off the next development cycle in 2007 by introducing the iPhone, which sounded the death knell for the desktop computer revolution that he had done so much to bring about. This time the copycats weren't IBM or Microsoft but Google with its smartphone operating system Android. Together, Apple's iPhone and Google's Android phones created the mobile boom. Thanks to their marvelous touchscreens and internet connectivity, they quickly became the blockbuster products of the age, much to the chagrin of formerly dominant phone makers, such as Nokia, Blackberry and even Microsoft, who were left behind in the dust.

The smartphone and its larger cousin the tablet have already become powerful enough to challenge the dominance of the classic PC. While personal computers may have brought the internet into our homes, it was the smartphone that brought us all together in one ubiquitous online community.

The milestone of five billion smartphones worldwide is already in sight. Smartphones create a more intense and personal digital experience than ever before. The present imperative dictates that everything that can be done via smartphone will be. Contrariwise, anything that doesn't happen via the smartphone often doesn't happen at all.

The third cycle of digitization is causing more and more aspects of everyday life to be governed by the web. The digital → customer journey is becoming the path most followed. And that doesn't just mean communication, entertainment, information and purchasing, the web touches our lives in more subtle ways as well. The average user reaches for their smartphone at least 150 times a day as we use these digital moments to avail ourselves of microservices tailored and presented to us personally.

The smartphone has become a way to remote control the world, to call up every service imaginable. Every company should therefore concern itself with finding its own button on the remote, thus turning the established value chain on its head by offering each and every customer a convincing value proposition. The spread of the smartphone has not only been a big shot in the arm for e-commerce, it has also forced the other three marketing Ps (product, price, promotion) to rethink themselves.

It might seem self-evident but it still catches many vendors by surprise that the internet has led to complete price transparency, greatly increasing the pressure on many traditional businesses who still don't seem to get it. Advertising (promotion) also needs to reinvent itself. The kind of interruptive advertising that worked so well for more than 150 years from the very first newspapers in the middle of the 19th century to magazines, movies, radio and, finally, TV is rapidly becoming obsolete. Digital users don't allow themselves to be interrupted unless the interruption is entertaining.

Modern consumers are also increasingly resistant to classical forms of advertising. Ad blocker software is becoming popular all over the world. Many consumers today will only accept advertising if it is perceived as highly relevant and personally tailored to their individual needs. This is where the real advantage lies for companies like Google, Apple, Facebook and Amazon – known collectively under the acronym → GAFA – because of the wealth of personal data they control. Still, the auction model on which they are based poses a major loss of efficiency, resulting in a highly dysfunctional system of advertising for many advertisers, as we will discuss later.

For the product, the final P in marketing, all this poses a huge challenge. The existing tools like advertising (promotion), distribution (place) and financial modelling (price) offer little room for differentiation.

Cycle #4: IoT and AI (2020-2025)

In the summer of 2016, Apple's management did an interview with Fast Company magazine in which they predicted the iPhone would continue to have a good run over the next 10 years or so. Perhaps we really won't find anything better than the smartphone till then – but it isn't certain.

Over the past three S-curves, computer screens have become smaller and smaller. It used to be that computer monitors were as big as TV sets. Then notebooks with much smaller screens made their way from the study to the kitchen table or onto our laps. With the advent of the smartphone, screens became even smaller. The next step will probably not be a tiny smartwatch but products with no screen at all! Voice systems and personal assistants like Google Assistant, Apple's Siri, Microsoft's Cortana and Amazon's Alexa no longer need a visual interface.

This development may very well herald the next S-curve with the new generation of devices being controlled through voice recognition systems or, as in Internet of Things (IoT) applications, through gestures and sensors. As for when this will happen, the first three S-curves offer some clues. The initial cycle (office PCs) lasted 20 years, the second (web PCs) 15 and smartphones only 10. This could mean it will only take five years for

→ artificial intelligence (AI) systems with voice and gesture control to take over. If true, we could very well be standing on the threshold of the next cycle which could last from 2020 to 2025. This scenario presupposes that the human voice will soon replace visual interfaces.

But back to the present. We are currently seeing the explosive growth of mobile services causing the web to expand to fill even more aspects of everyday life. Presumably the Cambrian Explosion more than 500 million years ago caused its own S-curve, with the first primitive building blocks of life being supplanted by more sophisticated life forms, leading to an explosion of new species thanks to the chance effects of mutation. Finally, saturation was reached as most ecological openings were filled.

The building blocks of the digital ecosystem consist of cloud services, mobile (LTE or 4G) telecommunications and stationary (WiFi) networks, special mobile processors (ARM, Apple, Intel), the smartphone operating systems iOS and Android, and, at the very top, touch-sensitive apps. Combining these building blocks in novel ways opens a plethora of opportunities, but it does contain risks. This is especially true at the cutting edge between hardware and software – the so-called Internet of Things – where a wealth of new products is being created that tends to increase complexity rather than reduce it.

One of the best examples is the Philips Hue system that supposedly enables customers to control their home lighting systems remotely through an easy-to-use app. Regardless of whether finding, starting and using a smartphone app is actually easier than flipping an old-fashioned light switch, customers report darkened homes because of fluctuations in the WiFi signal and, in some cases, it has proved impossible to dim bedroom lights because the owner has neglected to update the security settings for the switches and lamps.

This may sound amusing but there is a more serious context here. In October 2016, the biggest → distributed denial-of-service (DDoS) attack was unleashed against the basic infrastructure of the internet. In America, the services of NetFlix, PayPal, Spotify, Twitter and many more companies became unreachable or were greatly diminished. What had happened? An anonymous group of hackers had gained access to millions of IoT devices through security loopholes in internet-linked products like light switches, toys, baby monitors and surveillance cameras. These were used to create a "zombie army" of transmitting devices capable of flooding the servers of major internet providers with millions of queries per second, causing them to crash.

Developing innovative services within a complex ecosystem, it turns out, is anything but trivial. Interfaces, security and user-friendliness all call for a high degree of expert know-how. Many old-school manufacturers lack this expertise because it's simply not part of their DNA – unlike most software companies. In fact, the ability to develop the right software to bring secure, value-adding services to existing hardware will become a key marketing differential – the secret sauce of any new product.

The rise
of the GAFA

PLATFORM ECOSYSTEMS

Each of the three preceding cycles in personal computing was determined by a dominant → platform. The word itself is a bedazzling term that conjures up a number of meanings, often leading to confusion in discussions about them. Sangeet Paul Choudary, co-author of the book *Platform Revolution* (2016), has developed an architectural framework that distinguishes between three platform layers:

Network/Marketplace/Community Layer. At this level users interact either directly through social networks, or indirectly by exchanging goods or services (marketplaces). Some platforms contain an implicit community level, like Nest's learning thermostat or Mint.com's personal finance manager, that combine data from many customers to create added value for these users without them having to collaborate directly with each other. Some networks run by external service providers like software developers (App Store) or property owners (Airbnb) operate at this level.

Technology Infrastructure Layer. This level concerns the technology behind the platforms. It makes no sense if users and partners can't add value to a platform. External participants need to be able to build on it, such as apps in the smartphone world. The infrastructure level can be quite substantial – the Microsoft Windows and Intel (Wintel) alliance, Apple's iOS operating system, and Google Android – or only relatively skimpy (Instagram). As platforms grow the initial scarcity of users and partners tends to flip and become a problem of superabundance instead. This growing complexity threatens to make the platform irrelevant. How do you find the best offers, products or services? The answer is simple: through data.

Data Layer. Every platform contains a data level. This can be more or less pronounced but data sets create relevance by matching valuable content, goods or services with the right users. There are some platforms whose significance depends solely on data.

Following Choudary, we can now explore the different platform configurations within this framework:

①   Platforms like Airbnb and Uber, the darlings of the sharing economy, as well as networks like Facebook or Reddit which are based on the network/marketplace/community level. For these, the network itself is the main source of value creation.

②   Developer platforms like iOS and Android where the infrastructure level is crucial. Value is added here with apps in combination with the overlying network/marketplace/community level.

③   The third group of platforms is dominated by data. Instances include wearables, IoT and the Industrial Internet (sometimes, especially in Europe, referred to as Industry 4.0).

Data availability is poised to provide the impetus for the fourth wave of personal computing marked by intelligent personal assistants and human-machine-interaction through gestures and voice due to recent advances in AI. The speed of their appearance came as kind of a surprise since they were based on decades-old scientific findings in areas such as neural networks and → deep learning.

This involves algorithms capable of learning from real data in multitiered networks. They rely on the immense amounts of data made available thanks to Big Data. Large internet platforms are essentially just massive data aggregators, and today, for the first time in the history of AI, the volume of data has reached the critical mass necessary to train algorithms – yet another → unfair advantage for the GAFA companies under current market conditions.

What are the drivers behind the predominance of the GAFA platforms? At most, the history of Google, Apple, Facebook and Amazon only goes back about four decades.

THE FIRST PLATFORMS

My very first computer was a Commodore VIC 20 back in 1981. Commodore was founded during the 1950s in New York by Jack Tramiel, the guy who went on to save Atari from bankruptcy when the videogame boom went south. His Atari ST/TT was a really low-cost alternative to the ruinously expensive Apple Macintosh. Tramiel was a visionary who landed one blockbuster hit after another, each regularly reaching seven-digit sales figures. In Germany, where I lived, a professional developer scene sprang up around the Atari ST/TT, including the Signum word processor, the Calamus publishing tool and dozens of MIDI interface applications for music makers.

Tramiel had his first big success with the VIC 20, which Commodore billed in Germany as the "VolksComputer" (the people's computer) and named it the VC 20. One of the great ironies of fate is that Jack Tramiel enjoyed his greatest successes at CeBIT, the huge computer fair held every year in Hanover – just a few miles away from the concentration camp at Hanover-Ahlem where he, as a 16-year-old inmate, was forced to make tires for another Volks product 40 years earlier – the Volkswagen car popularly called the VW beetle.

Having moved to the US, Tramiel launched his career after World War II importing typewriters and producing pocket calculators. By undercutting rivals like IBM and Hewlett-Packard he managed to build up an impressive market share. He followed this strategy with PCs, too, generating huge sales volumes until he could offer them at a price every household could afford. In many ways, Jack Tramiel was the complete opposite of Steve Jobs. His computers were made in Asia, came in cheap plastic casings, and were run by chips Tramiel took on responsibility for manufacturing for himself so as not to be beholden to the big chip makers like Intel and Motorola.

But in the early 1990s it became apparent that his greatest hits – the VIC 20, the Commodore C64, and the Atari ST/TT – would have no follow-ups. Tramiel had become trapped in a world of his own making, one in which products were developed and sold as separate lines and where competitors were pushed aside through massive marketing campaigns and cut-throat pricing. Over time, the platform model transformed the PC market, and Commodore and Atari, along with other early pioneers like Sinclair and Texas Instruments, slowly withered away and died.

THE WINTEL IMPERIUM

In the mid 1980s, Microsoft and Intel got together to transform the PC market into a platform business. Together with IBM, which played the role of the hapless midwife, they created the platform that was to dominate the PC market for more than two decades. The strategy they followed can still be seen today as the blueprint for establishing platform-based business models.

While IBM gave its name to the IBM PC Standard, major components of the platform were not in fact controlled by Big Blue, as employees fondly nicknamed the company. For one, the young Bill Gates managed to do a deal with IBM for his MS-DOS operating system that did not give IBM exclusive rights. Microsoft was free to license MS-DOS to any manufacturer they liked. Even the reference architecture for the IBM PC came from yet another third party – Intel, a tiny upstart chip maker. Together, these two companies would eventually establish the famous Wintel alliance, one of history's most successful platform businesses.

Other manufacturers could use Intel's reference architecture and chips to create computers that were 100 percent compatible with the genuine IBM PC – which was lots more expensive. The result was a bitter price war between Asian and American PC makers. This, along with direct distribution (with Dell leading the way) and the emergence of big discounters like Best Buy in

the US and Vobis in Germany, as well as the continued fall in the price of computer chips eventually made Bill Gates' dream of a "computer in every home" come true.

Thanks to the hardware platform's openness a gigantic third market arose, leading to the spread of new applications which enlarged the platform's scale. As IBM clones became more popular and standardized, creating software for the platform became increasingly attractive for independent developers. The greater the number and range of software products became, the more attractive PCs became for potential users. This virtuous circle is the secret of good platforms: Success begets success. Once a platform reaches a certain critical mass, liftoff is achieved.

Creating a successful platform involved lots of hard, behind-the-scenes work. Microsoft needed to treat the partners in its ecosystem equally and orchestrate their often-diverging interests. They provided powerful tools for programmers and, together with Intel, created standards and drivers which ensured that the hardware and software of an army of vendors and developers would continue to work together seamlessly. By improving the PC's → usability they made it even more popular. Finally, the introduction of the Windows operating system opened the door to the mass market. Each new player – programmers, hardware makers, users – added yet more value to the platform.

By the time Windows 95 came along the duopoly of Microsoft and Intel was so firmly established that the entire PC market had become part of the Wintel platform. IBM finally decided to leave the market, Apple suffered an existential crisis and only barely managed to hang on in a niche market. Even the attempt by Steve Jobs, who had been kicked out of Apple, to establish an alternative computer platform called NeXT failed at first.

NETSCAPE AND THE WEB

The triumph of the Wintel Alliance and its platform in the mid-1990s carried with it the seeds of its own decline into irrelevance. In 1993, Marc Andreessen, a student at the University of Illinois at Urbana-Champaign, invented the first web browser with integrated graphics which he called Mosaic. Only two years later, Netscape Communications, the company Andreessen founded after graduating, was taken public in one of the most-watched IPOs in stock market history. The same year Microsoft introduced Windows 95 which lacked internet capability and had to be hastily updated. Bill Gates, who reputedly dismissed the internet as "a passing fad", finally came around to realize that Netscape posed a huge threat to the highly profitable Wintel platform.

Initially, the internet didn't belong to PCs at all, it led a quiet existence among university students and their ilk and ran on the type of Unix workstation popular there. It was Tim Berners-Lee, a young scientist at the European nuclear research facility CERN in Geneva, who used his NeXT workstation to create the "World Wide Web" and develop one of its basic concepts, namely the HyperText Markup Language (HTML), as well as the first primitive browsers and web servers.

The example of his own success had taught Gates that markets can be rolled up from below by a technically inferior product. That was the reason IBM took its own product, the PC, so lightly in the beginning – and probably why its team negotiated such a lousy contract with Microsoft. At the time, big mainframes and powerful departmental minicomputers running Unix or VAX marked the state of the art, and that's what Big Blue had in mind when it thought about "serious" computing.

Microsoft was afraid the web would eventually become the operating system for a new generation of network computers. At the same time, its other cash cow MS Office came under threat. Technically, they worried that a web browser would do everything that Excel or Word did. In reality, the web was much too slow back then and it took 20 years for broadband to make that vision possible – like this text, which was written in Google Docs and not in Microsoft Word.

The open architecture of the web exposed yet another line of attack. Netscape had the potential to create a new common layer on top of Windows, MacOS and Unix and so effectively establish a new standard which would one day marginalize the Wintel platform.

Gates reacted by drying up all of Netscape's revenue streams. Microsoft's browser was free for all users, its web server was optimized for the Windows browser and sold to customers under very attractive licensing agreements. This would later cost Microsoft a fortune in antitrust fines but at least it saved the platform and Netscape was neutralized. Despite this, it wasn't Microsoft that came to dominate the web but the GAFA quartet consisting of Google, Apple, Facebook and Amazon.

GOOGLE

While the browser wars of the 1990s were being fought between Microsoft and Netscape, two Stanford students Sergey Brin and Larry Page were thinking long and hard about how to search the internet. They wanted to create a search engine that would bring order to the chaos of the web. Early search services like Yahoo were curated by humans and just couldn't keep up with the explosive growth of web content. Computerized engines like Lycos, Infoseek and AltaVista provided more hits but were unable to weigh results according to relevance, limiting their usefulness. Brin and Page developed an algorithm that was capable of organizing web content according to a system they called PageRank (PR). Each web page was graded according to the number of incoming and outgoing links related to it. The original Google paper stated:

We assume page A has pages T1...Tn which point to it (i.e., are citations). The parameter d is a damping factor which can be set between 0 and 1. We usually set d to 0.85. C(A) is defined as the number of links going out of page A. The PageRank of a page A is given as follows:

PR(A) = (1-d) + d (PR(T1)/C(T1) + ... + PR(Tn)/C(Tn))

At the beginning, realizing their theories was costing Stanford so much bandwidth that college officials told Brin and Page they would need to move their nascent company off campus. In 1998, Google was born. The secret behind the search engine's stellar success wasn't just the search technology itself but something that deserved to be ranked as the digital product of the decade: AdWords. This is an online advertising service where advertisers pay to display brief advertising copy to web users. Nowadays, AdWords hauls in more than a billion dollars per month and keeps Google's coffers full to bursting.

When the first paid AdWords advertisement appeared on the Google search page in October 2000 the internet had just lost a lot of its luster. The last startups had managed to sneak onto the Nasdaq before the sky fell. The Dotcom Gold Rush had run its course and the bubble burst. Just half a decade before a few intrepid venture capitalists had started to place bets on fledgling internet companies. The trickle became a stream and then a flood. Everybody was convinced that the internet had created a whole new economic model. Driven by rapidly falling transaction costs and buoyed by expected network effects, natural monopolies would be established in virtually every sector of business. The Swedish pop group ABBA provided the anthem for the age: The Winner Takes It All. Punters doubled and tripled their stakes, a few PowerPoint decks were enough to prove that this latest techie dream child would be the Next Big One, with exponential profits virtually guaranteed.

Most proved to be pipedreams, and when they crashed against the solid wall of reality they went up in smoke. Most internet novices failed to find efficient ways to reach new customers, and instead they burned their IPO earnings by investing in ever more extravagant marketing ploys. TV stations and even print media reaped a bonanza selling spots and ad placements but most of the lucre wound up in online channels, especially banner ads. The boom fed off the boom, online ads were bought and sold just like they were back in the days of the very first advertising agencies in the 1850s. Banner ads were priced per view at fixed rates and with the usual discounts.

Ad prices were oriented around the same metrics used in print advertising. Around the turn of the century the cost per thousand, or cost per mille rate (CPM), averaged about 50 dollars. But things didn't add up. Given an average click rate on an online ad of 0.5 percent, advertisers were seeing costs of anywhere north of 10 dollars per individual visitor to the website. If only four percent of these clicking visitors eventually became paying customers, the cost of acquiring a single customer could be as high as 250 dollars or even more. This was unsustainable for most enterprises, and by the end of the year 2000 it was pretty clear to most people that the web lacked a viable business model.

Google's invention of AdWords came just in time to save the web. Brin and Page needed a way to monetize their search engine as it hewed its pathway through an ever-expanding web. The number of queries was increasing daily, and the two founders were being swamped by demands for additional infrastructure. Since they lacked a sales team of their own, they decided to create a do-it-yourself advertising system that customers (or their agencies) could manage on their own.

Brin and Page also agreed that they didn't want to sell format ads but instead only to show text ads if they corresponded to certain keywords, hence AdWords, that matched the advertiser's context. In addition, clients were expected to pay, but only if someone clicked on the link to their page (a system now known as "pay-per-click", or PPC). The price of a certain AdWord was to be determined by auction, with the highest bidder ranking first according to the laws of supply and demand.

This ingenious mechanism creates a link between the interests of the user and matching advertisements from Google's clients. The users divulge their intent via the search box – what's he or she looking for? What is engaging them? Google supplies advertising that is highly relevant in the context of the search through an algorithm, and it does so at a price that has been determined in an open market by auction bidding, all of which happens in real time.

In this model there are no intermediaries, no middle men selling ad space and bundling ads from different clients to persuade vendors into giving a pocketable discount. A student with a prepaid debit card can set up shop purchasing AdWords just as easily as the biggest advertising agencies, both getting the same conditions from Google. The beauty of this is that Brin and Page always achieve the best possible price for each AdWord they sell because advertisers are letting them know exactly how much a click is worth to them.

The moment Google AdWords entered the scene, the internet was instantly transformed into the most efficient marketing platform the world has ever known. At first, advertisers were only willing to spend peanuts per click, because only a handful of first movers were willing to shift their spending towards Google, and this kept auction prices low. As long as the price per click stayed at 50 cents, big e-commerce vendors like Amazon who were able to convert around five percent of their visitors into sales were paying the equivalent of maybe 10 dollars per new customer – a pittance compared with pre-Google days.

Companies of all kinds and sizes began to test the business model, initially with rather limited budgets, and lots of experimenting went on to optimize spend. No more risky, costly ad campaigns: the entry barriers were lowered dramatically. That was crucially important because many internet companies who had suffered from the collapse of the dotcom bubble no longer had the wherewithal for lavish ad spending: They needed to finance their marketing out of their cash flow, essentially spending as they went along. Thanks to Google, every cent spent on marketing could now be exactly correlated with turnover. The cost/sales ratio became the primary performance measure for internet companies in virtually every sector from retail to tourism, telecommunications to finance.

Google took the web and transformed it into a platform inhabited by users and an ever-growing number of powerful e-commerce companies, both linked together by a web of searches and AdWords. When Brin and Page started out to bring order to the web, they could have hardly imagined that within a few years they would have risen to become the central ordering force for e-commerce. Not only marketing became reliant on Google: entire business models now depend on the company, along with product ranges, pricing and user experience.

Google's auction model puts pressure on the margins of advertising clients. For many, the internet is now their most important interface with their customers, and marketing and sales budgets have shifted to reflect that by focusing primarily on digital channels. In the auction arena, this creates a situation that was first advanced in game theory as the Prisoner's Dilemma. If a company wants to achieve an edge over another it will need to keep raising its bids for AdWords, which reduces the margins. If, on the other hand, the company chooses to increase their margin by reducing their bids, they pay a price through declining market share and revenue.

Corporations face the added problem that auctions have since come to dominate other online media markets as well. Sheryl Sandberg, who led Google's AdWords business unit from 2001, moved to competitor Facebook in 2007 and introduced the auction model there, too. Today, virtually the entire advertising business, not only at Google and Facebook but on most major platforms, is conducted via → real-time bidding. The shift to auctioning transformed the competitive arena into the proverbial rat race.

Google remains as the only member of GAFA committed to an open web. Presumably, the company's executives lose lots of sleep contemplating a fenced-in web of the future, consisting of separated app silos whose content Google's web crawlers can't penetrate and index. The company's future lies in the world of mobile, and it's no coincidence that Google's current CEO Sundar Pichai used to head the Android unit responsible for the company's mobile phone operating system. Pichai's favorite project is called Accelerated Mobile Pages (AMP) and is dedicated to creating a new kind of user experience at the native app level in order to preserve the web as a (Google) platform. Google was among the very first to create truly Transformational Products including such blockbusters as Google Search, AdWords and Google Apps. Another significant product is Google Maps, which deserves a closer look:

- **Google Maps transforms user expectations.** With Maps, Google has dramatically changed what users expect from this product category in a number of ways. For instance, the continual upgrading of map details and Street View images, and allowing users to donate their own data, such as photos, opening hours, etc., to update the maps in real time has added new dimensions. Google also constantly adds new layers to its map service such as current traffic conditions or alternative transport modes, thus increasing its value to customers.

- **Google Maps transforms user behavior.** By personalizing maps and integrating other Google services (like Mail and Calendar), Maps is becoming an everyday companion for many users. Google Assistant and the OK Google voice interface, for instance, now help to inform users ahead of time about any traffic gridlocks and ways to avoid them, so they can still arrive in time for dates or meetings.

∞ **Google Maps transforms how value is added.** Google Maps is by far the best map service in the world and doesn't cost its users a cent. Google makes money by seamlessly integrating ads into its maps. Maps is now a strategic product for Google which gives it access to car users and manufacturers, thus beefing up its Android ecosystem. With its free service Google has seriously curtailed the market for standalone satellite navigation aids like TomTom or Garmin, and the services put pressure on automakers to integrate Google's service into their models. As self-driving cars enter the market over the next few years Google will be in an ideal position to cash in on millions of additional hours users spend with its service.

AMAZON

Like Google has done with its AdWords auctioning system, Amazon is impacting the margins vendors can achieve online. Its founder, chairman and CEO Jeff Bezos has been quoted as claiming that "Your margin is my opportunity." In the two decades of its existence, Amazon has cut into more and more product segments and hence itself has become a platform. Marketplace was Amazon's first big step in this direction. As of late 2016, more than 70,000 dealers were active in Marketplace.

The dealers fulfill a variety of functions for Amazon but, basically, they act like truffle pigs to nose out attractive new ranges of goods and products. The minute a dealer's product starts gaining traction and selling in volume, Amazon can consider introducing it to its own lineup. This causes other vendors to follow suit, and prices start to spiral lower and lower. Once the scale effect sets in, Amazon can inflame the price war if it wishes and thus optimize its own margin.

If a product is substitutable, Amazon can create its own private brand, which allows it to sell at an even lower price, putting pressure not only on vendors but also on manufacturers. Amazon currently sells more than 3,000 products under its own label. In the United States, 44 percent of all customers say they routinely search for products on Amazon versus going to a search engine like Google or navigating a manufacturer's website. This gives Amazon reliable data about overall market demand which enables it to fine-tune its own lineup and pricing.

A look at Amazon's turnover suggests that the company is currently on the front edge of the S-curve and will continue to grow. In fact, it seems that sales at Amazon are continuing to increase. Second quarter results in 2016 topped Christmas sales of the previous year for the very first time. In his annual statement to stockholders, Bezos still greets readers with the optimistic message: "It's still day one."

The company's sales curve seems to support his chutzpah. Amazon's expansion rate has given Bezos the opportunity to establish his very own e-commerce ecosystem that now looks almost unassailable. By investing billions in physical infrastructure Amazon, unlike Google, has built an additional firewall around its business model.

Amazon's physical logistics capabilities have become an essential part of the business strategy of many third-party manufacturers and vendors. This strengthens Amazon's own business model and gives it an edge over pure-play digital enterprises. Yet, for all that, Amazon remains a software company at heart. Its worldwide logistics network from its hubs all the way to the buyer's front door is run by an ingenious set of software tools developed and constantly upgraded by Amazon's own software engineers. Thanks to Amazon Web Services (AWS), Bezos' company has become the biggest cloud service provider in the world, side-sweeping the established players like IBM, Oracle and Microsoft in the process.

In addition, Amazon is hard at work establishing a physical presence in the homes of its customers. Products like Fire TV, Echo and the Dash Button system allow it to place a foot in the door of millions of customers around the globe. Even more important, personalized services offered by Amazon are being upgraded through massive use of AI. Bezos is not only building the biggest sales platform in the world, he also wants to provide the best errand service for physical and digital products. Therein lies the crucial difference.

Prime, the customer loyalty program Bezos has created, must not be underesti-mated, neither in the short term nor in its impact on the future of his company. In fact, it was one of Amazon's very first Transformational Products.

Prime transforms user expectations. Thanks to Prime, Bezos is cre-ating constantly rising customer demands. Amazon will deliver any product that carries a barcode free of charge to your home, reliably and in increasingly short order. In the language of perceptional psychology, this is known as → priming. Thanks to the scale effect of both its abil-ity to deliver and the densification of its logistics, Amazon's customer experience has become both unique and virtually unassailable. Amazon's speed and reliability are the yardstick against which users define their expectations in the realm of e-commerce.

Prime transforms user behavior. The former online bookstore from Seattle is busy adding more and more services to its loyalty program, things like Prime Music, Prime Video, Pantry, Same-day Delivery, Kindle Book Loaning, Prime Photos and Twitch Prime. Instead of simply priming their customers' psyche, Amazon is achieving real customer → lock-in – the dream of every vendor. Products are effectively pre-sold as soon as the customer clicks over to the Amazon website. Market researchers at Millward Brown have found that the probability of a Prime member buying something when they visit the site is an unbelievable 74 percent. That is more than 20 times the average for a normal online shop and proves that Bezos is anything but a normal online dealer.

Prime transforms how value is added. Just as Jack Tramiel was the antithesis of the young Steve Jobs in the 1980s, Amazon today is the antithesis of Apple. It is successful in e-commerce because it doesn't play by the rules. Bezos isn't out to make the best tablet or the best e-book reader. Atoms are only a (necessary) manifestation of their service. More than anything, Amazon is an errand service provider.

FACEBOOK

Facebook is a product of the Age of Mobility. Its newsfeed and messenger services are now used almost exclusively on mobile devices. With his acquisition of mobile-only players like Instagram and WhatsApp, founder Mark Zuckerberg has shown a deep understanding of what it is that "Generation Touch" wants. As a result, Facebook's earnings from mobile advertising are going through the roof. By single-mindedly focusing on enhancing his portfolio by following the lead of Chinese competitors, such as Tencent with its "super app" WeChat, Zuckerberg is apparently grabbing the chance to increase the clout of his platform significantly.

Today, the glue between Facebook and its users is still the newsfeed. Originally this consisted for the most part of personal posts from friends and acquaintances. Today, Facebook's real-time algorithms curate a personalized stream of integrated news and ads that seems just like the kind of personal newspaper developers dreamed of back in the early days of the web. Younger users especially use Facebook as their main source of news. The wiping gesture to refresh and scroll through the newsfeed is strangely similar to pulling the handle on a slot machine. Both serve to satisfy our craving for distraction, novelty and amazement. Surprisingly, booming advertisement videos form relaxing islands of peace in this ceaseless attention-demanding torrent.

Facebook's future is being decided not in Silicon Valley but in China, which is leading the way into the Mobile Age. No other CEO is as focused on China as Zuckerberg, who, since 2014, has been a frequent visitor and has been astonishing local businessmen with his growing command of the Mandarin language. It isn't the sheer size of this potential market that fascinates "Zuck", it's the digital parallel world in Asia's biggest country that interests him most. China's → walled garden insularity protects its digital pioneers in more ways than one. On the one hand, there is the so-called → Great Firewall of China with which the government effectively censors the internet by blocking certain IP addresses and through → deep packet inspection of data. On the other, a number of closed platforms like Tencent/WeChat and Alibaba have sprung up and become highly successful, effectively blocking US platforms from getting a foothold in the country. Chinese users effectively leapfrogged the first two waves of personal computing. For them, the internet is all about smartphones and apps. WeChat, especially, resembles AOL back in the 1980s in some ways. Every → use case you can imagine, from content to entertainment with e-commerce, communications and chats, are all found within a single self-contained application.

Encapsulated platforms like WeChat are the blueprint Facebook is seeking to follow as it moves forward. Having cornered the market for mobile advertising thanks to the attention magnet of its newsfeed, Facebook is now looking to turn Zuckerberg's Messenger product into an intermediary. An increasing share of the estimated five billion dollars Facebook spends on research and development every year is now being funneled into AI projects. Smart bots, they hope, will one day make users' everyday lives easier and more convenient. Consumers will one day be able to do shopping, ticket purchases, reservation bookings and more via Facebook's Messenger – using text or voice.

APPLE

Steve Jobs and Steve Wozniak founded Apple, the eldest member of GAFA which is also the only one to have created two successful → pivots over the course of its comparatively long history. Without the breakthrough of the Apple II in 1977 personal computing and the web (which was developed on Job's NeXT workstations) might never have happened, just as the mobile internet owes its existence to the iPhone singularity that materialized in 2007.

Google, Amazon and Facebook are essentially digital companies for whom the physical is just the means to an end of binding users to their platforms. They view the world through the eyes of software developers, going so far as to open source their innovative hardware free of charge. Apple does the opposite, binding users to their physical products through digital services. The contrast couldn't be greater, and it makes Apple's business model less reliant on user data for its monetization. In its contrary world, Apple actively encourages its third-party developer community to create ad blockers and anti-tracking tools. This serves the dual purpose of positioning Apple in the eyes of their customers as a credible data protectionist while, at the same time, undermining the data-based business models of Google and Facebook.

It was the introduction of the iPod music player that kicked off Apple's mobile hardware-centric platform philosophy that has now made it the world's most valuable company within the space of a mere 10 years. Tony Fadell, who later founded Nest and sold it to Google, was a contractor at Apple and pitched the concept of a business model that combined both hardware and a service at the company's Cupertino headquarters in the year 2000. It was his idea to create a closed ecosystem consisting of hardware – a music player – and software for organizing and purchasing music. Apple not only took the idea and created the iPod player and iTunes store, but in the process transformed the entire music industry. This was thanks to an ingenious negotiation ploy by Steve Jobs. By emphasizing Apple's relatively tiny share of PC sales (the Mac held less than five percent of the market in 2001) he tricked all the major labels into licensing their catalogs to his iTunes Music Store for a mere pittance.

Up to that point all attempts by the music companies to tackle the problem of rising music piracy had failed and Apple looked to them like the ideal test case due to its puny size. However, within 18 months Apple began to offer its free iTunes software for Windows PCs, too. The music player sales exploded and by 2005 the iPod had become Apple's most important product. Buoyed by their success, Apple went on two years later to introduce the iPhone. In the meantime, the iTunes online store had become the biggest distribution channel the music industry had ever known.

Even the iPod bore all the attributes of a Transformational Product:

∝ **The iPod transformed user expectations.** iTunes software, the iTunes Music Store and the iPod formed a single unit and synchronized among themselves. For the first time, Apple managed to bring the ease of use of Mac programs to consumer electronics, where engineers for decades had more or less ignored the user interface. That way, Jobs gave consumers a totally new and lasting user experience that surpassed even that of Sony – at that time Steve Job's prime model.

The iPod transformed user behavior. The original iPod came with a sensational, at that time, 5GB hard disc that Toshiba had only recently launched onto the market in an ultra-compact 1.8 inch format. This gave the iPod the capacity to store about 1,000 songs. Users began to transfer their entire collections of CDs to the device and went on to buy new titles and whole albums directly from the iTunes store. The Compact Disc, which was introduced in the mid-1980s and had pushed record sales to one new height after another, began to wither with the advent of the iPod, to finally be ousted from the minds of many consumers – and finally from shopping bags and department store shelves.

The iPod transforms how value is added. Apple had undergone a near-death experience in the middle of the 1990s and subsequently brought back their founder Steve Jobs who they had infamously fired just a few years previously during another crisis. Within a year of his triumphant return Jobs managed to stabilize the company's core business by introducing semi-translucent iMacs in a range of colors. The "i" stood for the internet but it could just as well have stood for Jonathan Ive, the brilliant designer who, for the next 20 years, was to form the counterpart to Jobs, creating a new and innovative design language for Apple products. It was the iPod that truly transformed the former PC pioneer. With it, Apple created a totally new product category, set itself at the head of the music industry and turned itself into a pop icon. The iPod was the Trojan horse Apple used to → hack the lost Wintel market, and its hybrid business model was to become Apple's blueprint for the future. Five years later, Apple would go on, thanks to the iPhone, to become not only the most successful company in the world but also to effectively reinvent personal computing once again. By 2016, Apple's revenues from services (apps, cloud services, music and video) was growing faster than its hardware business for the very first time.

CHINA: ALIBABA AND WECHAT

We have seen how the personal computer evolved from the geeky gadget of the 1980s to become the mobile, ubiquitous daily companion of today. Google, Apple, Facebook and Amazon used this physical manifestation to create a series of hugely successful services that now form the interface between companies and consumers. In this chapter, we have also discovered just how big a role these gatekeepers play in monetizing that interface. The GAFA members have become the most valuable corporations in economic history and together are worth more in the stock markets than the GDP of South Korea. While the duopoly of Microsoft and Intel merely dominated the IT industry in the 1990s, today's crop of techno giants dominate the whole economy.

GAFA's Transformational Products show that value in the digital world is still being created at the interface with the customer. This axiom is reinforced if we look at the largely sealed-off Chinese economy. 90 percent of online retail in China is conducted over platforms. The websites of individual

From important to dominant

When Tech was scared of Microsoft, it was a pretty small company

GAFA
WINTEL
ALL OTHER

Top 20 US listed companies by market cap (indexed) – 1995

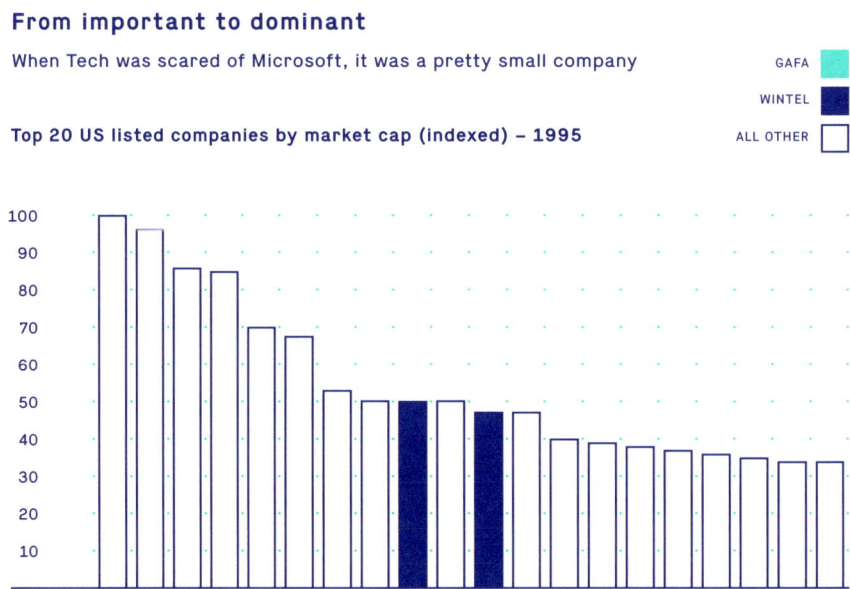

vendors and manufacturers play only an insignificant role. The single largest marketplace, Tmall by Alibaba, encompasses more than 150,000 retailers and 200,000 brands. Tmall charges vendors an entrance fee and gets a cut from every transaction conducted through it. On the other hand, it offers its partners high visibility, lots of traffic, an easy way to manage their offerings and an interface that is optimized for mobile users. Even companies like Apple turn to Tmall as an alternative to building hundreds or even thousands of retail stores in this huge country. Its market cap of around $380 billion makes Alibaba one of the most valuable companies in the world.

WeChat, another Chinese company, plays in approximately the same league in terms of valuation. Many Chinese use this service to organize numerous aspects of their everyday lives. In a single app, WeChat combines commerce, payment, social, mobile, local and, through its → application programming interfaces (APIs), a wide range of additional services. The company's genesis is unique. Although it already owned QQ, one of China's most popular messaging services, the mother company Tencent decided to create an entirely new

Top 20 US listed companies by market cap (indexed) – 2016

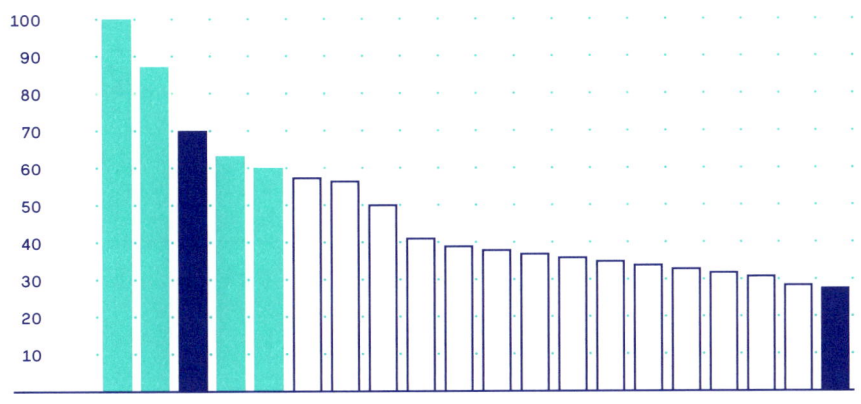

Fig. 3: Market capitalization of the top 20 US listed companies

messaging service that would focus entirely on the mobile market, which it subsequently expanded to become a fully-fledged platform.

WeChat is the quintessential Transformational Product. Depending on your viewpoint it can be a mighty platform, the front end for a slew of services or the service layer for a plethora of products. As a platform, WeChat is a match for those provided by GAFA. As a service front end it resembles Google and Facebook by functioning as the interface between vendors and customers, thus competing in the customer relationship arena. WeChat goes a step further by functioning as the service layer for the products themselves which subsequently become interchangeable. Need a taxi? WeChat can order one for you. WeChat ingests mobility services like Uber or Mytaxi and makes them obsolete. As a messaging platform, WeChat plans to do the same to a wide range of mobile business models.

The first part of this book has shown how personal computing has progressed over the past decades in three waves by shifting value creation to the user and gaining control over the customer interface. In the second part, we will delve deeper into the question of what products need to look like if they are to wrest back control over that interface in a world of ubiquitous networked hardware, platforms and services.

CODE

Transformational Products

Transformational Products have three core features: They redefine users' expectations, change their habits, and alter the way companies create value.

Transformation of user expectations. Products should offer outstanding perceived benefits that change what users expect from the product category as a whole – true transformation rather than mere incremental improvement.

Transformation of user behavior. Innovative products lead to lasting changes in the way the product is used. Marketing is often tightly integrated with the product itself.

Transformation of how value is added. The product has the potential to significantly boost turnover and revenue and/or safeguard an existing business model. Transformational Products seldom face resistance within the organization and are widely greeted as catalysts of change.

In this chapter, we will explore how these three elements combine as illustrated by the **EXPERIENCE LOOP**, but first we must discuss a fundamental property shared by all Transformational Products. Essentially, they all consist of a bunch of digital services that beget unforeseen usefulness, sometimes in combination with a physical component.

SERVICES ARE EATING THE WORLD

"People don't want to buy a quarter-inch drill. They want a quarter-inch hole!"

– Theodore Levitt, economist

"Software is eating the world", Netscape's founder Marc Andreessen claimed back in 2011, thus instantly coining one of Silicon Valley's most lasting modern memes, the kind that bridges the gap between insight and self-fulfilling prophecy. With this one he nailed down the difference between a hardware-based economy and one based on software – one of the major paradigm shifts of our time.

That's only part of the message the phrase conveys. At least as important is the fact software-enabled services are putting pressure on the physical world. We have marketing gurus Stephen Vargo and Robert Lusch to thank for this insight by which they described the shift in focus from an economy based on goods to one based on services.

In 2004, they published *Evolving to a New Dominant Logic for Marketing* in the *Journal of Marketing*. The article advanced the concept of service-dominant (S-D) logic as a meta-theoretical framework for explaining value creation through exchange. According to their reasoning, services are responsible for the greater part of the value added by a product. Goods, they argued, are just a distribution mechanism for service provision. S-D logic leads to the realization that the main contribution of companies producing goods is to orchestrate a complex bundle of services between various actors and institutions, such as capital owners, designers, suppliers, factories, logistics partners, marketing departments, retailers, and so on. The physical proxy (for instance, a car) only masks the services that led to it.

Writing more than a decade ago, Vargo and Lusch didn't explicitly mention digitization in their work since their intention was to devise a general theory of economics. However, this is exactly where the explosive power of their argument lies: Services are governed by software logic and can be scaled exponentially at negligible cost regardless of the amount of human effort involved in its creation. The human service is transformed by software into a digital service. Wherever humans remain indispensable for performing the service (think taxi drivers, pickers in warehouses, or administrative assistants) they are only condoned as intermediate steps in establishing the digital service itself.

S-D logic describes yet another important aspect of digitization: The value of a physical product, such as a car, lies not in its constituent atoms, that is to say its material share in value creation, but in its utility; a property Vargo and Lusch call its → value in use. This is distinct from its → value in exchange. For instance, Spotify as a media streaming service may have great practical value for its users, but it has no exchange value.

The authors ultimately introduced a concept they called → co-creation which states that the value of a product is co-created by multiple actors which always includes the user or, in goods-dominant terms, the consumer. An electric drill only benefits its user when it is used. The real value of the product is therefore determined by its value in use.

In the digital world, value is created by the users as, over time, they personalize their own search history, for instance, by returning to Google time and again, and it is the user who clicks on the AdWord advertisement. Thus, the user takes on the role of → co-creator, allowing self-learning AI systems from Google to continually improve an individual's results by analyzing their search behavior.

The authors also call our attention to the fact that this effect isn't really all that new. It was just never as explicitly obvious or so clearly understood as it is today. The IKEA home goods acquisition concept, for instance, only works when the customer is an active participant. When you have the bookshelf delivered to your home and assembled there by a courier firm, the service you're paying for costs a lot more than the particleboard from which it is made. Transport and assembly become an inherent part of the product itself. In fact,

IKEA has been struggling for years to adapt its business model to the digital world because these services cannot be reproduced digitally.

A product is something that has been manufactured, so it is material in nature, but it is also the direct result of services performed by human beings – and increasingly by software. In addition, the physical products themselves are increasingly being augmented and enriched by the service layer. Without iTunes and the ecosystem Apple has built around it the iPod would have remained nothing more than a rather expensive portable hard drive.

It's also apps and cloud services that have transformed a lowly vehicle like a Smart automobile into the linchpin of the trendy metropolitan car-sharing service Car2Go. It therefore makes sense to focus on the digital service layer when developing a new product because the laws from that realm offer an easy way to increase the usage value of physical products in traditional services' markets by adding network effects, scalability and exponential cost benefits – as we have already seen in the first part of this book.

The success of a product is increasingly dependent on the services it offers its owners. The same goes for entire enterprises. Banks and insurance companies are in full retreat in the physical world, as more parts of their businesses turn digital. Other industries, such as telecoms, are in the process of closing their data centers and transferring these services to third-party cloud infrastructures. Even proprietary mobile phone cell towers and receiving equipment are disappearing because it doesn't make sense any more to build your own when it's easier to share them with other providers like municipal utility companies. Here, again, the physical side of the enterprise is being virtualized, and its products are becoming services. But beware the hidden trap, as the American venture capitalist Benedict Evans remarked:

"It is easier for software to enter other industries than for other industries to hire software people."

The bundle of services orchestrated by the enterprise is clearly emerging as the major value-add of products. This applies to every kind of product – even automobiles. Cars may be the quintessential physical product, but the services surrounding them are becoming the key distinguishing factor. How does the car fit in with our other daily services like messaging, entertainment or navigation? What are some of the other services you can enjoy while sitting in your car? Do you even need a car of your own anymore or could you join a car sharing service, instead?

It remains to be seen just how massive the effects of the service layer will be on cars as products but it does seem certain car engines will become increasingly commoditized thanks to electrification. At the same time, the expectations of car owners will increasingly be shaped by the user experience of digital services outside the car. Rupert Stadler, CEO of Audi, is quoted as believing that his company will one day generate more than 50 percent of its revenues through digital services, in effect achieving a balance between the physical and the non-physical aspects of the automobile.

For Transformational Products, digital services are the key value driver. The world of atoms, S-D logic tells us, is really just a proxy, an expedient way to package services. It is like a token we pass on to prove that a service has switched owners. Today, we would probably prefer to say: "Services, not software, are eating the world."

THE DISCOVERY OF USEFULNESS

"Users do not care about what is inside the box, as long as the box does what they need done."

– Jef Raskin, Apple GUI designer

Every successful product solves a real problem. What S-D logic describes as "value in use", Clayton Christensen, the author of *The Innovator's Solution* (2003), calls "jobs to be done". People choose products because they want to solve a certain problem. Christensen illustrates this through the example of a fast-food chain selling milk shakes.

Traditionally, marketeers would split people identified as potential milkshake drinkers into groups following a variety of psycho-demographic faultlines. Then they would quiz their focus groups about preferences: Viscous or creamy? Chunky or smooth? Maybe just simply cheap? This gives clear results but it's hardly a good way to create new products that will boost sales or profits. If that's what you want, you have to take a completely different approach by asking consumers how milkshakes impact their lives. For example, research shows that almost half of all milkshakes are purchased and consumed in the morning.

Further questioning shows that most of these customers are commuters looking for a way to brighten up the dreary drive to work. They aren't really hungry when they buy their drink, but they know that by about 10 o'clock they will be. They're in a hurry, they're wearing office clothes, and they usually only have one hand free. Bagels leave lots of crumbs and give you greasy fingers. A banana is consumed too quickly, so it can't dispel the boredom of a long commute. A milkshake is the perfect answer!

This may sound trivial, but it's insights like these that give product teams a clear idea of how they can improve a product and where, in the eyes of the customers, their true competition lies. The use value of a milkshake may be very different from what you and I would think intuitively. To make a better milkshake, you need to uncover its hidden use value.

Transformational Products, it turns out, need an additional attribute besides the three we have already discussed: At heart, they're just a bunch of digital services that happen to come with an optional physical component. In the next section, we will see how these attributes combine to create the **EXPERIENCE LOOP**.

EXPERIENCE LOOP

Digital transformation isn't driven by enterprises themselves. The change vector starts with the user, who is also the co-creator of the added value. The value shared between users and enterprises is the transmission belt that moves from **SERVICE DIFFUSION** to **SERVICE EXPERIENCE**, and then on to **SERVICE CO-CREATION**. In our model, we call this conveyor belt the **EXPERIENCE LOOP**.

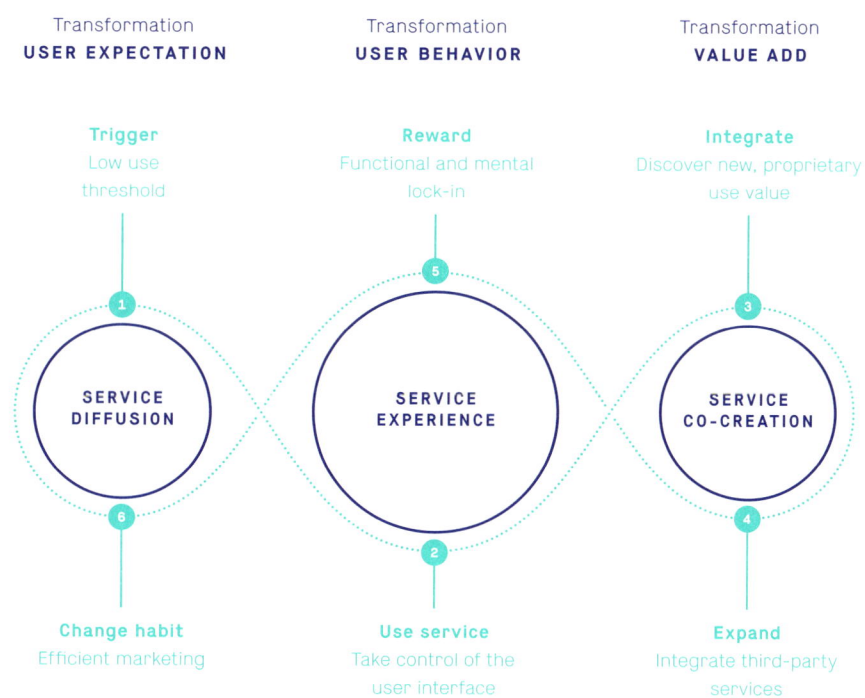

Fig. 4: Experience Loop

Transformation
USER EXPECTATION

Transformation
USER BEHAVIOR

Transformation
VALUE ADD

Trigger
Low use
threshold

Reward
Functional and mental
lock-in

Integrate
Discover new, proprietary
use value

SERVICE
DIFFUSION

SERVICE
EXPERIENCE

SERVICE
CO-CREATION

Change habit
Efficient marketing

Use service
Take control of the
user interface

Expand
Integrate third-party
services

The six steps of the **EXPERIENCE LOOP** can be illustrated using the example of a chauffeuring service:

1. **Trigger – low use-threshold.** Easy onboarding via an app with the offer of a free ride as an incentive to sign up.
2. **Use service – take control of the user interface.** No instruction is necessary to use the interface; vehicles are tracked and driver ratings add a degree of transparency traditional taxi companies can't hope to match.
3. **Integrate – discover new, proprietary use value.** Create a decentralized taxi network and use network effects to bring passengers and drivers together more efficiently. The bigger the network the more value it brings to both drivers and passengers (number of rides, waiting time, vehicle availability).
4. **Expand – integrate third-party services.** Use external providers for login, tracking, payments and navigation services. Integrate offers for car sharing, shared rides, and public transport.
5. **Reward – go for functional and mental lock-in.** Cars should be available at very short notice, the drivers well-qualified. Being more transparent than other car services and offering more convenient payment methods, this service provides a better – and, above all, constantly reliable – user experience than conventional taxi companies. The app makes calling for a cab unnecessary.
6. **Change habit – efficient marketing.** Create a tight bond between customer and service through mental and functional lock-in to reduce customer acquisition costs over the entire product lifecycle. Satisfied customers can be incentivized to recommend the service to others.

The **EXPERIENCE LOOP** spans three stations which we will now explore in more depth.

SERVICE DIFFUSION – transforming user expectations

Transformational Products change what users expect from an entire product category by fulfilling a **radically different value proposition**. This always starts with the customer who is first lured and subsequently convinced by the product's ease of use **(casualness)**. Transformational Products offer significantly greater benefits that are highly relevant to the consumer's everyday life and are therefore perceived as more than simply an incremental improvement. In short, they do the job at least an order of magnitude better, for instance 10 times faster, cheaper or more conveniently **(10x value)**. The value thus created will often induce users to share their experience with others. Finally, using the service frequently involves non-users. Transformational Products are inherently viral. These products are therefore often self-promoting **(built-in marketing)**.

SERVICE EXPERIENCE – transforming user behavior

Ideally, customers will become habituated to using the service as often as possible, thus creating value for both the enterprise and themselves. Transformational Products should always offer a central login from which users can personalize and reconfigure the services provided. The process of snapping the product into the life of the customer is called **achieving lock-in**. This can be a **mental lock-in**, which happens when users begin to perform certain tasks automatically. **Functional lock-in** occurs when interacting with the digital product through the **user interface (UI)** becomes a habit. The involvement and interaction between the customer and the product is called the **user experience (UX)**.

SERVICE CO-CREATION – transforming the value add

Transformational Products interfere with traditional value chains and force companies to rethink their **business model**. The new one needs to have the potential to create significantly more revenue and profits than the old one, or at least to reinforce an existing business model. In both cases **scale** is crucial. The best way to integrate additional third-party offerings is through application programming interfaces **(APIs)**. Data points are important for personalization (especially in conjunction with artificial intelligence and provide value of their own when combined with additional third-party services **(data)**.

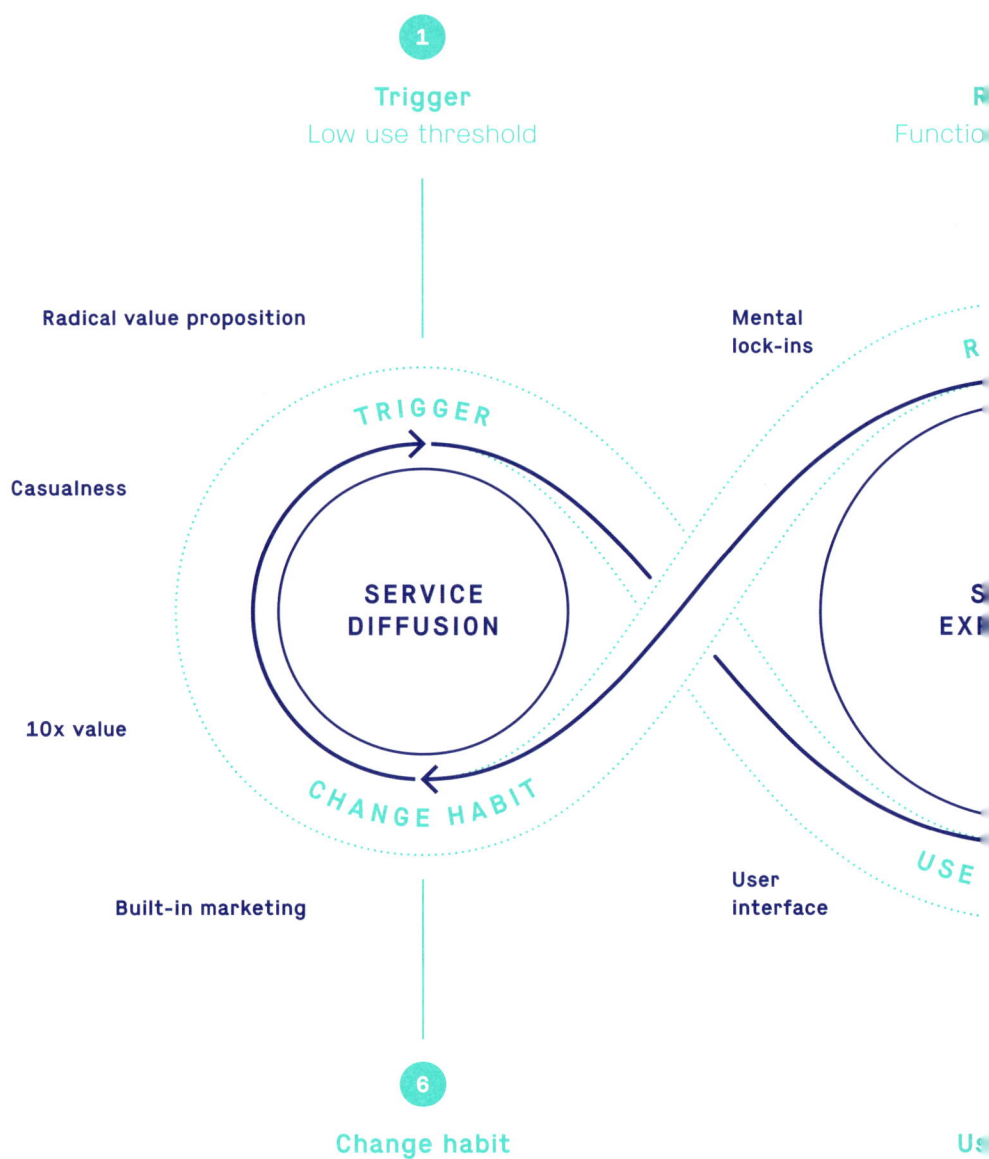

Transformation
USER EXPECTATION

Tran
USER

1

Trigger
Low use threshold

R
Functio

Radical value proposition

Mental
lock-ins

R

TRIGGER

Casualness

**SERVICE
DIFFUSION**

S
EX

10x value

CHANGE HABIT

USE

Built-in marketing

User
interface

6

Change habit
Efficient marketing

Us
Take
use

Transformation
VALUE ADD

3

Integrate
Discover new, proprietary
use value

mental

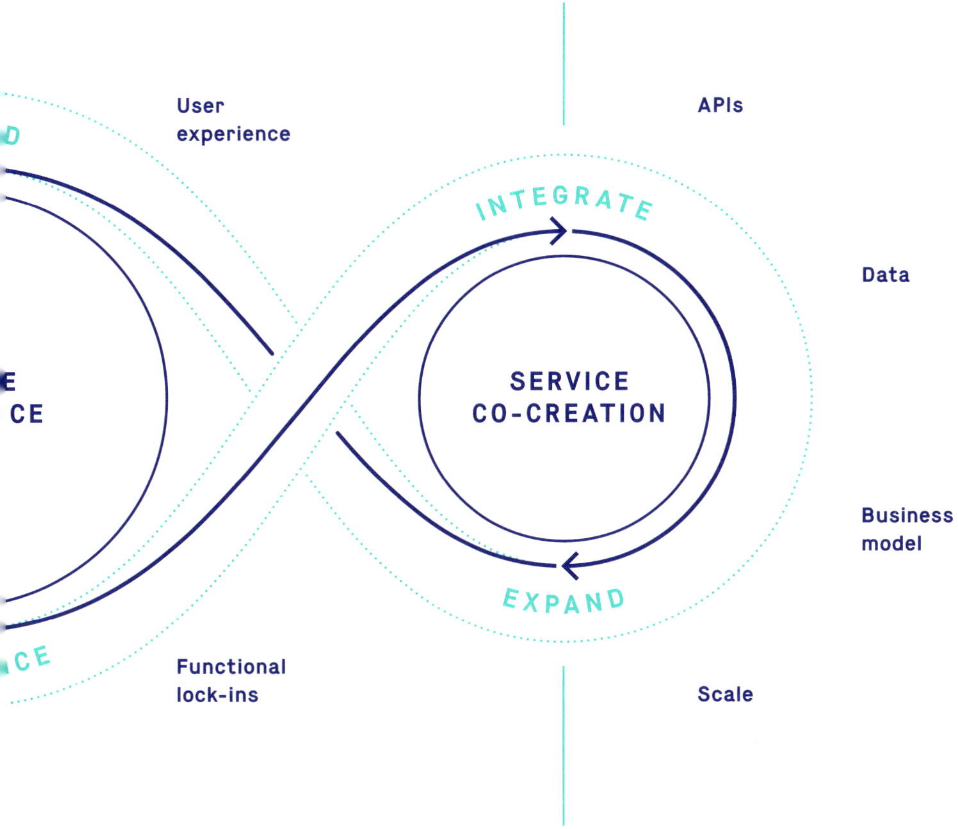

**User
experience**

APIs

INTEGRATE

Data

**SERVICE
CO-CREATION**

E
CE

**Business
model**

EXPAND

**Functional
lock-ins**

Scale

CE

4

Expand
Integrate third-party services

ice
of the
ace

SERVICE DIFFUSION – transforming user expectations

Truly innovative products mold markets by transforming what customers expect from the whole product category. Amazon's 1-Click shopping is a good example which, together with Amazon Prime, has completely changed most people's view of e-commerce. Interestingly, changes in user expectations can cross borders into other product categories, too. Uber, for example, is following a strategy in the realm of mobility that is strikingly similar to Amazon's. Uber offers a service that will get you a taxi fast, cheap and reliably with just one click. Another example might be Amazon's Dash Button concept that automatically orders everyday products in the physical world, like detergents or diapers, when they run out.

On an abstract level, the examples shown above are about fulfilling needs instantly and well (on demand). Not so long ago it was big brand names that offered a sense of security in deciding what to buy, now it's digital services that perform this function. They move in and occupy the space in our brains where buying habits are formed, as well as occupying the corresponding buttons on our smartphones. In the old days of analog, a taxi service needed a telephone number that was easy to remember but, now, all Uber or Mytaxi need is their own, easy-to-use app. This opens up a huge opportunity for Transformational Products to discover the problems facing, and the hidden needs of, people and turn them into products that change what we all expect from the product category.

Transformation
USER EXPECTATION

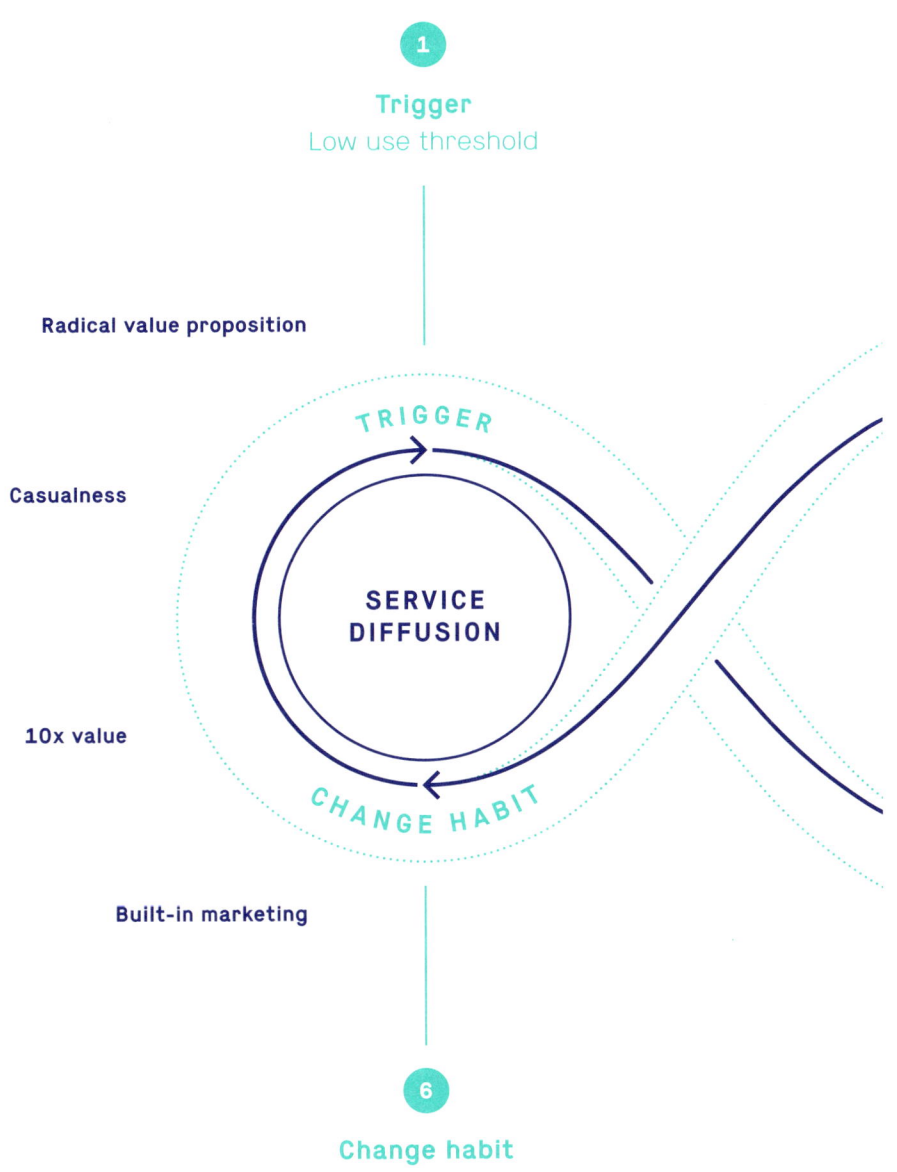

1

Trigger
Low use threshold

Radical value proposition

TRIGGER

Casualness

SERVICE
DIFFUSION

10x value

CHANGE HABIT

Built-in marketing

6

Change habit
Efficient marketing

CASUALNESS

Transformational Products are to classical software products what casual games are to complicated strategy games. They need to be designed to function without a mediator and without demanding too much mental gymnastics from the user. They are like casual games that provide instant entertainment at hardly any effort. Designing Transformational Products is user-focused and requires a deep understanding of the user. Traditional product design, of course, works the other way round, starting with legacy systems and business processes and ending with the user.

The concept of an "end user" is relatively new in IT, first appearing in the 1980s. It dates from the rise of the PC during the first wave of personal computing. In the two preceding decades people who used computers in enterprises were almost always experts. This monopoly of the geeks came to a close thanks to the revolution in microelectronics which put computing power in the hands of the rest of us – something we today refer to as the → consumerization of computers. The term is popularly ascribed to Douglas Neal and John Taylor at IT research firm Leading Edge Forum in 2001. It denotes the merging of enterprise and consumer which used to be strictly separated. People have become used to taking their gadgets to work and using them professionally (the "bring your own device" concept) and using their company equipment after hours in the "home office". After all, they can access the company systems from anywhere they want thanks to the ubiquity of the internet. In some companies, this goes so far there are no longer any real offices for employees to go to, or at least the employee no longer has a desk of their own in a world of "non-territorial" workspaces or "hot desking."

For IT professionals, this revolution has forced a new way of thinking and working. Like it or not, they were compelled to take the user into account, to make themselves more approachable and to create products that targeted the user. Developers and users are dragged (sometimes kicking and screaming) into a partnership that aims to benefit them both. This effect led to the rise in the 1990s of what is now called → user-centric design. Over the course of the last two decades this approach has permeated through various schools of design, all sharing a common virtue, namely the goal of optimizing usability in new products instead of, as in the past, forcing behavioral change on the user to fit in with the product.

In Transformational Products, this change of focus is taken to the extreme. The functionality of a product now needs to be completely intuitive. Quite often, users aren't even required to make conscious decisions and can nonchalantly play around to discover the benefits of a new service which then unobtrusively finds its way into their everyday life. The bond between the product and the customer can best be described as unconstrained and → casual.

RADICAL USE VALUE

Successful companies are generally quite good at getting things right. With Transformational Products, the trick lies more in delivering the right things as quickly as possible – and not just promising to deliver. The time span available for enterprises to make hay by making improvements to a success-ful product has become severely limited. This seems funny because Silicon Valley and Seattle used to be famous for their grandiose visions, also known as → vaporware. For instance, Bill Gates announced as early as 1994 his vision of "information at your fingertips". However, the founder of Microsoft didn't believe in the internet at all, he wanted people to use his proprietary Microsoft Network. He eventually discovered the error of his ways and tried to correct his course, but in the end, it was his new rival Google that would hijack this dream. Brin and Page set out to organize all the data in the world and make it accessible to everyone but, unlike Microsoft, they didn't promise anything, they just did it. In the process, they delivered a product with a crucial edge – and that's what counts in the Digital Age.

Transformational Products promise a radically enhanced use value, and they deliver on their promises quickly. By providing rapid results, they create an immediate feeling of success, a real lift that's the first step towards lasting changes of user's habits. If this feeling is repeated it creates a positive experience which cements the product's position in the mind of the customer, triggering its increasingly routine use.

10X VALUE

Now as CEO of Google's mother company Alphabet, Larry Page isn't satisfied with simply improving an existing product by 10 percent. In his eyes, that's what everybody else is doing: incremental improvement. He wants to create products that are 10 times as good as the competition.

That's the difference between evolution and revolution, between linear and exponential change. You must completely understand what the problem is before you can rethink it and come up with a truly new solution. Gmail was created this way. Google's email service was started on April 1, 2004, and it was so radically different from existing email services many thought it was an April Fool's prank. While other providers were offering a miserly 20MB of storage, Google gave its customers 1GB from day one. You'll never have to delete emails again, Google promised. Instead, mails became searchable, just like the web. The front end was also unlike anything people had ever seen before. In some way, Gmail was less sophisticated than the industry standard of the time. For instance, it couldn't be used offline and required an internet connection at all times. However, its huge amount of storage space and the powerful search options offered were convincing arguments, and Gmail soon became firmly entrenched.

In 1997, management consultants Charles E. Lucier, Leslie H. Moeller and Raymond Held published a paper, *10x Value: The Engine Powering Long-Term Shareholder Returns*, arguing that the only way to create lasting shareholder value was by improving customer value by a factor of 10. Innovation, they believed, was the key to reaching tenfold improvement, which could be achieved either through strategy, products or services. A single "big idea" wouldn't be enough; you needed a whole slew of innovations.

Andy Grove, one of the founders and former CEO of Intel, identified a wide range of "10x Forces" in his blockbuster book *Only the Paranoid Survive* which he published in 1999. Each force was powerful enough to displace entire companies in the market: current and potential competitors, suppliers, customers, substitutors, and complementors. In other words, each component of a business model contains the seeds of its own disruption. In his book *The 10x*

Rule, Grant Cardone suggests that companies set their goals 10 times higher, and adjust their efforts and investments accordingly. That, he argues, makes the difference between success and failure.

This may sound like typical American business prose to you, but the point is that if you want to develop Transformational Products, there is simply no other way. Step-by-step improvement, Cardone maintains, just won't cut it if your aim is to lastingly change human behavior. If you slowly raise the temperature in the glass jar, the frog won't jump. Factor 10 change takes us to the verge of exponential transformation. It isn't enough for a product to be linearly better than the competition; the product and its individual components need to be dramatically better than previous solutions. If not, users probably won't jump and will stick with their old, familiar product.

This mindset is quite common among startups and venture capitalists. Michael Skok, a partner at Underscore VC and Entrepreneur in Residence at Harvard Business School, recently declared that "if you can't deliver a 10x promise, customers will typically default to ‚do nothing' rather than bearing the risk of working with a startup." Most risk-capital investors expect at least tenfold returns to compensate for the gambles they take. The same rule applies to Transformational Products, whether they are created by startups or more traditional companies. On the one hand, they must compete with startups seeking to increase their use value tenfold. On the other, they need to overcome the inertia of their customer for whom, after all, doing nothing is a real option.

BUILT-IN MARKETING

In 2010, entrepreneur Sean Ellis coined the phrase → growth hacker which contains two important elements: the focus on growth and using coding as a way to achieve it. Growth hackers understand both marketing and programming. They are able to integrate all aspects of modern marketing (branding, customer acquisition, onboarding, monetization, retention, virality) and make them part of the product itself. For them, growth has top priority, and we will see later why this is so crucial when we discuss scalability (→ scale).

TRADITIONAL MARKETING LEGACY

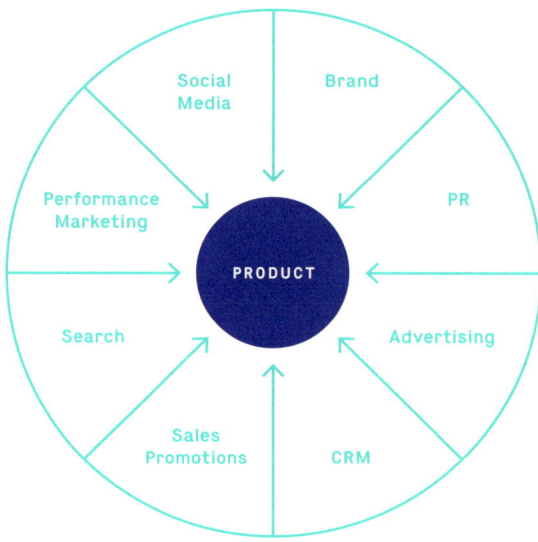

BUILT-IN MARKETING SERVICE LAYER

Fig. 5: Marketing legacy vs. built-in marketing

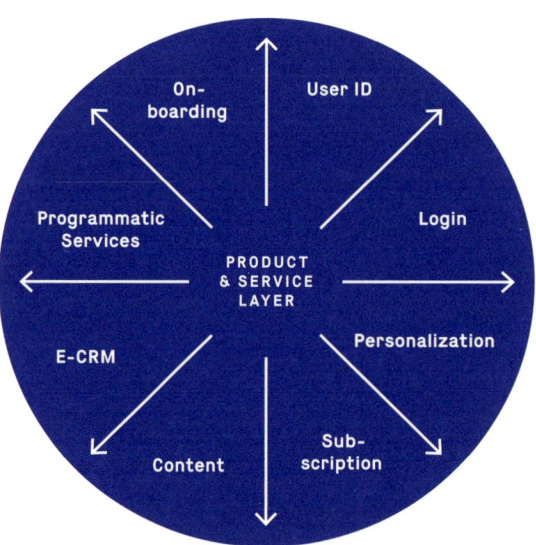

Making marketing part of the product illustrates an important philosophical shift. Traditional products carry a legacy from past efforts at marketing – brand identity, advertising, direct marketing, customer relationship management, public relations, events, channel marketing and many more. The first wave of digitization has added more categories – search, affiliate marketing, social media, and suchlike. External intermediaries such as media and sales organizations also belong in the legacy category.

Transformational Products already combine the product, the medium and the distribution channel: they are all three at the same time. The marketing legacy is replaced by a service layer that binds the user to the enterprise. The product doesn't need the stimulus of external marketing to persuade customers to use it. The whole point of Transformational Products is that their value increases with the frequency with which they are used.

Google, Amazon, and Facebook are among the world's most valuable brands today – and they never spend a dime on advertising. Their products possess an inherent use-value that sets them apart. Marketing performance is part and parcel of the product itself. Even Apple, a hybrid between a hardware manufacturer and a digital player, invests much less in advertising than its competitors, such as Samsung. In general, big spending on marketing is an indicator that a company has a problem in getting its product to sell.

Digital products, as a rule, depend far less on brand marketing than their physical counterparts. This shows itself in the offhand way most digital enterprises treat their brand logo. For them, trust isn't the result of a communicative process but of constant use. These companies don't need a brand by which people identify them. Like car sharing, digital products gain value by getting neighbors to join, too – resulting in better network quality and availability.

The best way to measure the self-marketing power of a product is by its so-called Net Promoter Score (NPS). The idea behind this is strikingly simple: customers are asked whether they would be willing to recommend the product to other users on a scale from 0-10. Anyone giving a score of 9 or 10 belongs to the extremely important group of promoters; users willing to give their unforced support to a product – true product evangelists. A score of 7 or 8 indicates a neutral, or passive, user. And a 6 means the individual would be likely to warn others against using the product.

SERVICE EXPERIENCE – transforming user behavior

In the old days of mass-produced physical goods, you only needed to sell a product once. Whether the customer ever used it or not was of no concern, nor did the manufacturer care whether it eventually became part of their everyday life. It therefore made sense to produce lots of standard goods and to market them massively by literally pushing them into the market. If enough customers bought the products, economic success was ensured. In the Industrial Age sustainability was a foreign concept. However, digital products don't work that way. A Transformational Product will only be successful if the customer uses it as frequently as possible (sustainability) and if it is tailored to him or her personally (personalization). The more it is used, the more value it creates. It is therefore necessary to change the user's habits.

LOCK-INS

In order to make a Transformational Product click into the customer's everyday experience (lock-in) you need something that sounds almost trivial, namely a login account. A user account as the basis for personalized services provides a way of achieving a sustainable relationship between the user and the product or service supplier.

Transformation
USER BEHAVIOR

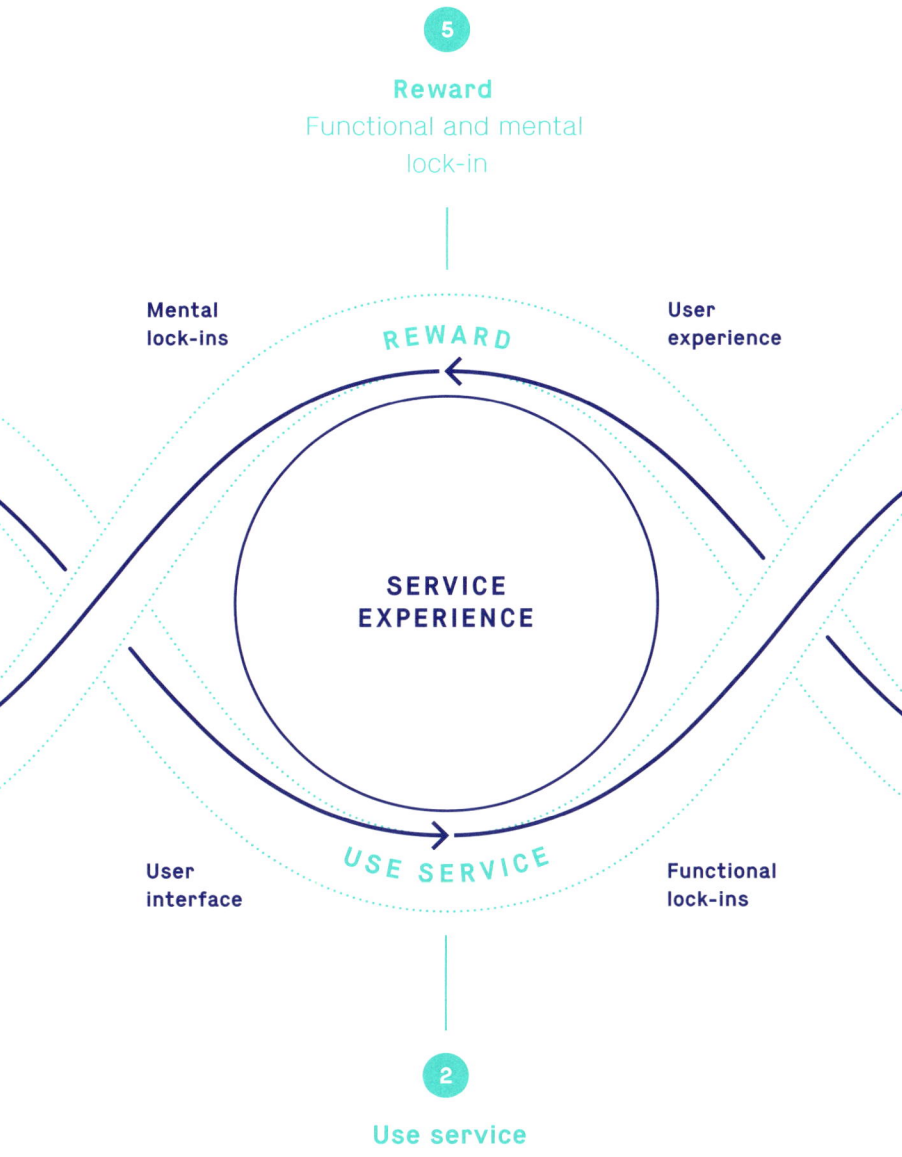

5

Reward
Functional and mental
lock-in

Mental
lock-ins

User
experience

REWARD

SERVICE
EXPERIENCE

USE SERVICE

User
interface

Functional
lock-ins

2

Use service
Take control of the
user interface

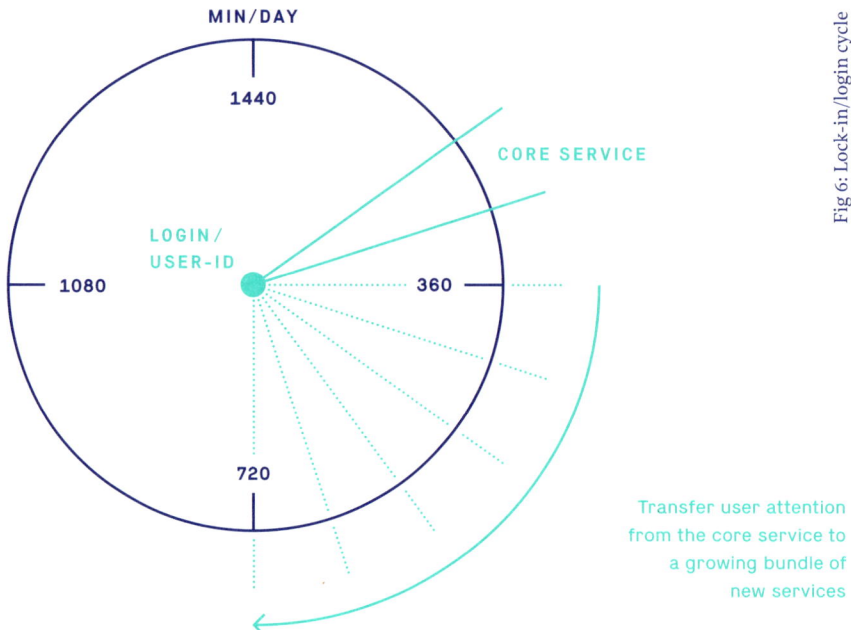

Fig 6: Lock-in/login cycle

MIN/DAY

1440

CORE SERVICE

LOGIN/
USER-ID

1080

360

720

Transfer user attention
from the core service to
a growing bundle of
new services

Sounds like a no-brainer? It's anything but, there are lots of counterexamples. Is the customer's phone number or email address key to an ongoing relationship with their internet provider? What about their utility company? Do insurance companies provide a central login for all types of policyholders? Are the personal settings of a car owner (like seat position, driving mode or navigation) linked to a personalized login or just held on the car key?

A login not only gives access to a person's profile and user history; it also has great strategic significance. It is the starting point for developing functional and mental lock-ins. It is the only way to use the company's core service, like search on Google or for ordering goods from Amazon, to attract the user's attention to a growing bundle of new services (see illustration).

Amazon, for instance, consolidates its lock-in with every new e-book a customer buys. The entire Kindle library is linked to the user's Amazon account, and they must log in to access their electronic books – a classic case of functional lock-in. The same goes for most other digital media libraries, as well as for email (Gmail), apps (App Store/Play Store), playlists (Spotify), or contacts (Facebook, LinkedIn, Xing). Even where there is no hard technological lock-in, so users could theoretically change providers at a whim without loss of data, convenience and routine usually lead to mental lock-in where there is no perceived benefit to the hassle and bother of switching to another system. It may be technically possible to transfer all your Kindle e-books to a different supplier's reader – if the Digital Rights Management (DRM) allows it – but why should you? The process would be perceived to be tricky and exhausting, and wouldn't improve the user experience much anyway. Therefore the average user would choose to stick with Amazon.

Subscription models are a perfect way to lock customers into a service. Customers of streaming services like Netflix or Spotify instantly lose access to new film releases and music titles the moment they cancel their subscription, and they get to kiss goodbye to their past investment in content, too. This → sunk cost fallacy leads them to cling mentally to their old product. In addition, changing providers would lead to the loss of the customer's user profile. But as we have seen before with the S-D concept, the user has been a co-creator and as such part of the value chain.

Sticking with the example of media libraries, users have not only been creating playlists, they have also been implicitly allowing the system's operator to anticipate their future preferences by analyzing behavior through powerful algorithms, and a constant stream of personalized recommendations will ideally have intensified the customer relationship. If they were to switch systems, users would forfeit all these benefits and essentially return to Go and have to go through the tedious job of wading through hundreds of irrelevant mainstream charts again.

The subscription model logic of digital products is nevertheless sharply different from classic subscription models, such as membership in a gym. Here, the best customers pay their dues on time every month without ever appearing for training. Or insurance: the best customers pay regularly without ever having to make use of services.

For business models like these, the mental lock-in achieved closely resembles the sale of indulgences in the Middle Ages which so enraged the reformer Martin Luther that he finally broke with the Catholic church and started the movement that led to Protestantism. Indulgences allowed penitents to bribe their way to heaven and were a kind of "sin now, pay later" scheme. But in the age of digital products, as we have seen, the marginal costs of a product trend towards zero. The supplier of a digital service profits every time his product is used because the usage costs are minimal and the revenue stream increases with use.

However, there are other forms of mental lock-in for us to consider, for instance "fire and forget". The past master of this art is without doubt Jeff Bezos of Amazon, who trains his customers to order stuff without thinking about it, preferably from a gadget supplied by Amazon. His hooking mechanism of choice is Prime, the customer retention program he dreamed up to provide free delivery in very short time with a high degree of reliability. An estimated 65 million customers worldwide use Prime, which generated over $6 billion for Amazon in 2016 alone. But Bezos still wasn't satisfied, and in December 2016 he opened the first bricks-and-mortar convenience store under the brand name Amazon Go.

Unlike traditional supermarkets, Amazon Go lets you walk in, grab products, and walk straight out without having to go through a checkout. All you need to do is check in to the store at the entrance by scanning the Amazon app on a sensor, and a range of technologies will make sure you pay the right amount for your shopping. They'll automatically know which products you've left the supermarket with, and charge you through your Amazon account afterwards.

Using Prime triggers a strong mental reaction that more or less eliminates all the major drawbacks of traditional forms of distance shopping, such as long waits, unpredictable delivery times and shipping costs. Bezos has once again upped the ante by introducing a same-day delivery facility in selected metropolitan areas around the world. The key to all this is 1-Click© ordering, a concept that Amazon pioneered and for which Bezos was rewarded a patent in 1999. It's another one of those charmingly simple ideas that leaves you asking why nobody had done it before. 1-Click combines concepts like trust, security, reliability, fulfillment, and speed and makes them part of the user interface, leading to even more intensive interaction on the part of the user. There is no better way to convey Amazon's promise of a unique user experience than this simple button on every product page.

Amazon Dash takes this idea one step further to achieve functional lock-in in the physical space. It adds a real button to everyday appliances and gadgets like washing machines or soap dispensers. By simply pressing the button, the user places an order for consumables with Amazon and gets the desired product within days or even hours. Once this idea has taken hold, the customer is soon "hooked" and will presumably want to add this convenient service to a growing number of appliances and consumables, taking Amazon far beyond the realm of the digital economy. One day soon, the manufacturers of coffee machines and dish washers will surely start to add this cute little button to their products before they leave the factory.

USER INTERFACE

The user interface is the intersection between humans and machines. Besides such familiar types of interface as websites or mobile apps there are a slew of others that allow users to interact with their systems. In the early days of the computer the most common interface was the command line where the operator had to type in complicated instructions for the computer to carry out.

Today, the link between users and their devices is the graphical user interface, or GUI. The idea sprang up in the early 1980s at the Xerox PARC Lab, where a group of engineers introduced it as a new way to control photocopying machines. Their brainchild, the Xerox Star, was the first graphical front-end to boast both windows and a computer mouse. Xerox later lost interest in their own idea for fear that it would cannibalize its core business. Bill Gates and Steve Jobs couldn't believe their luck when this once-in-a-century brainwave literally fell into their laps and quickly introduced it into their Windows and Macintosh operating systems.

Since this first Xerox moment the development of user interfaces has focused on making computers successively easier to operate. The complexity of the machine is increasingly being hidden behind new layers, all aiming at making technology more beneficial. Just as punched tape was replaced by terminals, windows and the computer mouse, this is now carrying over into touchscreen interfaces, gesture, and speech recognition and control. In the end, the interface will probably become an invisible envelope in which the user moves around. Everything will become causal.

"As far as the customer is concerned, the interface is the product", Jef Raskin believes, speaking as a pioneer in the field of human-machine interaction and one of the team at Apple responsible for the first Macintosh. Herein lies an important truth: users could not care less what lies behind the interface as long as the product does what they want. Competitors in a market could once rely on one-upmanship by constantly adding new features and bragging about them, but customers today are simply no longer interested.

During the first wave of personal computing, users could still enthuse about the vital statistics of their machines. Who, today, can get worked up comparing the specs of an iPhone versus an Android smartphone? In fact, smartphones are no longer distinguished by their hardware; the user experience is what makes the difference – in other words, the service it provides. Here, the user interface takes center stage.

Transformational Products closely align their services with the user's daily life. They don't stick out, or up, and they give the user a boost whenever one is needed. When everything revolved around the web, people would visit a company's homepage to perform a specific task. In the smartphone-dominated world of today, the service and its convenience will only intrude on the user's digital attention stream when necessary. The user-interfaces of Transformational Products no longer rely on flashy looks. Graphic design must focus instead on creating a system of interlocking components that cause and assist customer interaction. Colors, typography, and visual worlds aren't that important anymore. Instead it's all about a hierarchically structured design system like the one practiced by followers of the school of → Atomic Design which postulates that interfaces must do what they are intended for regardless, whether they be a two-inch wristwatch or an 80-inch home cinema system.

Uber and Amazon Go are harbingers of user interfaces to come. Both allow for casual use and settlement without importuning the customer. From a user viewpoint, the best interface could eventually be no interface at all. In any case, successful GUI design should aim at transcending the old rules in each product category and striving to change the user's routines.

USER EXPERIENCE

"You've got to start with the customer experience and work backwards to the technology."

– Steve Jobs

The secret of Apple's success in developing new products is to start with the user and the "user experience". In fact, the term can be traced back to Apple itself. In the 1980s, the company took Xerox's basic interface idea and pushed it forward. In doing so, the "Macintosh Human Interface Guidelines" were created in tandem with the machine's hardware, and this manual became the lodestone of a generation of GUI designers. However, Don Norman, another early Apple employee, recognized early on that concentrating on graphical interfaces would not be enough. Early in the 1990s he described a much broader concept:

"I invented the term [user experience] because I thought human interface and usability were too narrow. I wanted to cover all aspects of the person's experience with the system including industrial design, graphics, the interface, the physical interaction, and the manual."

Creating a unique user experience starts with the real user requirement.

Unfortunately, users are seldom able to articulate what they want or need. This must be uncovered and discovered before it can be translated into a distributed system of interactions between product and user. A good user experience subsumes several dimensions, including utility, usability, and → desirability, but in fact, without utility and usability, no desirability will follow.

⊙ **Utility.** The product must be able to solve an existing, relevant and re-occurring problem for the user or else satisfy a demand, and it must be able to do so better than anything the user has experienced with previous products, meaning 10x better, simpler, cheaper, and in a more fun way. If this is not the case, then you should definitely reconsider.

⊙ **Usability.** The user must be able to understand unequivocally and without lengthy explanation how to use the product and what it can do. Does the product hit the user's core need on the dot, and how easy is it to achieve the desired benefit? Of course this can vary depending on the target group and the user's prior knowledge. In any case, the product will miss its target and fail if the user is unable to understand what has to be done, regardless of the product's utility value, since customers will not use it unless they absolutely have to because there is no alternative.

⊙ **Desirability.** If a service is useful and easy to understand, it needs to be fun to use. If not, it will not become habitual. The more enjoyable the user experience, the greater the chance the customer will return to it again and again. Desirability also includes things like aesthetics, rewards, personal appeal, storytelling, but above all, product feeling.

Attention in developing user experience should always be given to the quality of each and every interaction. Developers are best advised to avoid the trap another popular method, known as → customer journey mapping, can lead to by focusing too much on analyzing the potential navigational pathways through various digital touchpoints. This always leads to an explosion of possible usage scenarios, thus squandering valuable resources without necessarily improving the quality of the product. Like Plato's cave allegory, thinking too much about possible permutations of the customer journey can soon become a shadow science. In developing Transformational Products, the focus must always remain on the true product experience. Functionality will never lead to habit change. The user's emotional product experience is at least as important.

To be seamless and comprehensive, a sustainable user experience requires both personalization and a consistent design system. As a general rule, usage trumps aesthetics every time. Amazon's user experience, for instance, consists of two parts. First, Amazon customers can find every product in the world that has a barcode, and if they can't find what they're looking for, it presumably doesn't exist – that's actually not completely true but it is, at least in the perception of most people. Second, anyone who orders something at Amazon will receive it the next day or even, at least in many metropolitan areas, the very same day or even within the hour. The parcel courier from DHL or Prime Now who delivers the product is part of the user experience and thus of the product itself.

The role of user interface (UI) and user experience (UX) can vary according to the product involved. The following table illustrates the difference for Amazon products and features.

	UI driven	UX driven
Functional lock-in	Amazon Dash	Amazon Kindle
Mental lock-in	1-Click Checkout	Same-day delivery

SERVICE CO-CREATION – transforming the value chain

According to the theory of service-dominant (S-D) logic, manufacturers of goods are also service companies. Seen this way, a car maker's value chain is essentially the way it orchestrates a complex bundle of services. It's not the end product that creates value, it's the way the company organizes its various subsections. Due to digitization, the use value of a product exceeds the value of owning it. Mobility is more important than the vehicle; the product itself becomes a service. As a service, it is governed by the laws of digitization: digits can be iterated much faster and cheaper than atoms. The digital service components of the product can be upgraded much quicker than the physical parts, so in the end they will contribute more value than the non-digital aspects of the product. Product development will shift more and more in the direction of software.

At the level of **SERVICE CO-CREATION**, the two parts of the value chain finally come together. The transmission belt that connects user and enterprise is what we call the **EXPERIENCE LOOP**. To function, this means that the user must get additional value and that there must be a viable and scalable business model for the product.

The co-creation principle also applies to the structures that create the organizational conditions for the product's scalability. Organizational transformation,

Transformation
VALUE ADD

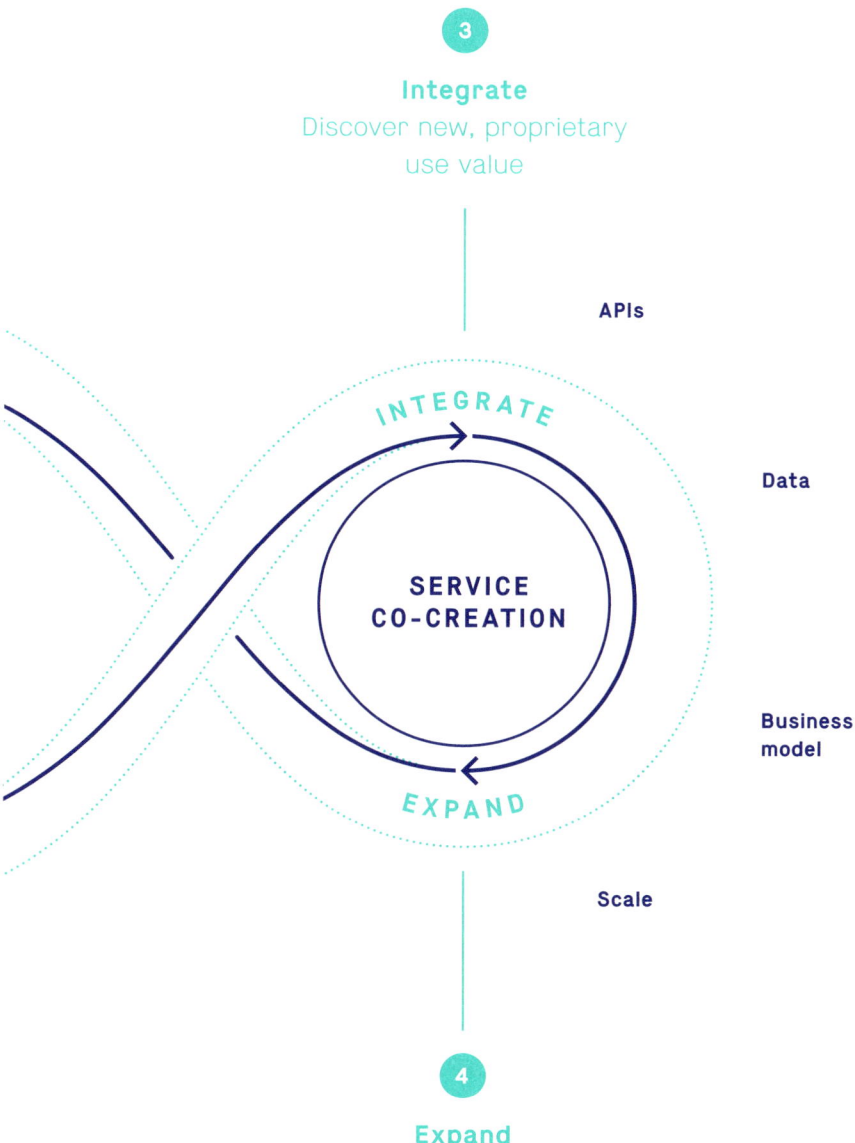

3

Integrate
Discover new, proprietary
use value

APIs

INTEGRATE

Data

SERVICE
CO-CREATION

Business
model

EXPAND

Scale

4

Expand
Integrate third-party services

however, always comes later. Transformational Products first change user habits and markets, only then do enterprises follow suit. Change in companies comes last, not first.

BUSINESS MODEL

The **EXPERIENCE LOOP** of Transformational Products follows a simple algorithm:

① Low-threshold usage
(SERVICE DIFFUSION)

② Capturing the user interface
(SERVICE EXPERIENCE)

③ Integration of proprietary and third-party services
(SERVICE CO-CREATION)

④ Create proprietary and new, or newly-discovered, value
(SERVICE CO-CREATION)

⑤ Create functional or mental lock-in for the customer
(SERVICE EXPERIENCE)

⑥ Efficient marketing
(SERVICE DIFFUSION)

⑦ Goto 1

This program can lead to any one of three possible business models:

Enrich & Defend

Here, the focus is on defending and securing an existing business model. The company's legacy products are expanded through a digital service layer, following Transformational Product logic, to improve accessibility, product experience, customer use value, diffusion power and customer lock-in. In this case the aim is not to increase monetization but instead to defend existing revenue streams. Since the new service layer at least partially supersedes old, traditional marketing methods adding it will not necessarily lead to additional costs. After all, the new digital services represent a direct and proprietary line

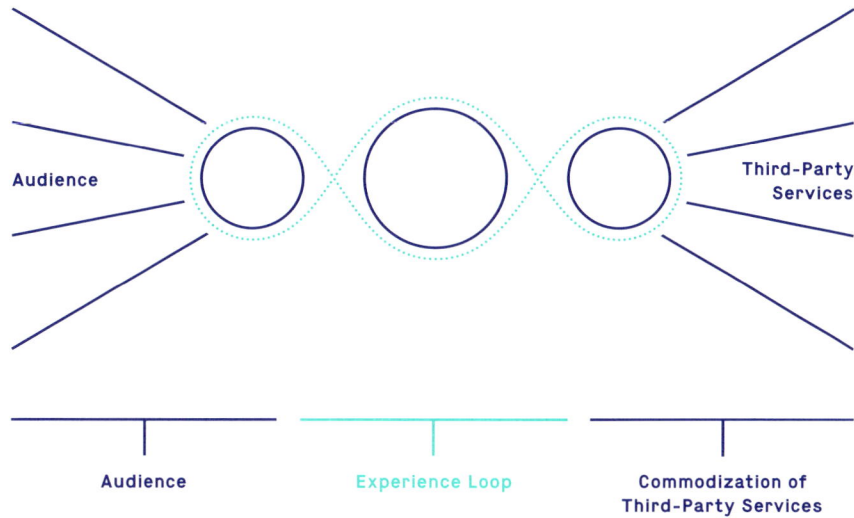

Fig. 7: Aggregation

to the digital customer, which promises not only a tighter relationship but also lower communication costs.

Transformational Products can therefore be seen as safeguarding, and in some cases, even leveraging the company's core business. At the very least they will stop the company from hemorrhaging revenue. A positive reverse effect on existing revenue models and assets such as brand, reach, sales channels, and in-house competencies is essential.

From the customer perspective, the Transformational Product becomes a service layer surrounding the old, established product which thereby seems more modern and appears to give higher use value due to its perceived proximity to the digital champions. In the best case, the service layer will keep the company from losing its customer interface, thus forming a firewall to protect the product.

Take "smart" online banking apps as an example. A large bank would be able to analyze individual withdrawals and payments by collating with its existing data to offer its customers a kind of automated household diary that gives them up-to-the-minute information about their financial status, along with a feeling of transparency and security about their account movements.

Create & Compose

This model is about creating new value, often through integrating third-party services. Here again, control of the user interface is key to cutting off third-party vendors from usurping the customer relationship, and in turn becoming commoditized. Along with expanding scale for one's own product the effect is to skim the margins of the integrated third-party services.

Thanks to digitization, indirect rather than direct revenues play a leading role. In extreme cases such as platform products from Google, Facebook and such, users pay nothing. In others, mechanisms such as free-to-play or pay-to-win borrowed from the world of online gaming are employed during the launch phase of a product or to bring new customers aboard an existing one. The first ride with Uber and the first month of Netflix are both free. So-called freemium models (derived from the words **free** and pre**mium**) generate turnover through upgrades to chargeable services.

Mix & Milk

Both models described above can be successfully combined. Apple, for instance, protects its own products through a robust layer of services (Apple-ID, iCloud). Simultaneously, it keeps creating new offerings built on these services (iTunes, App Store, Apple Pay) that regularly outperform the products from the core business. Established players should seek to emulate such hybrid models to both safeguard their core business and to find ways to create new markets. That's easier said than done, however. A dual-track strategy always leads to a race against time for established enterprises struggling with a legacy of antiquated systems and processes. The danger here comes from the digital champions who are hard at work trying to commoditize the existing products of virtually every industry.

Retail offers a cautionary tale here. As a procurement service, Amazon doesn't have to rely on the revenue it generates in sales. The rich streams of money flowing in from service products like Prime, marketplace commissions, and cloud services (AWS) are enough to enable Bezos to invest billions in his already powerful infrastructure, thus improving the customer experience step by step in ways that are far beyond anything his competitors can hope to achieve. Those unfortunates often have no choice but to adopt an "enrich & defend" strategy that relies on omnichannel initiatives, which in turn are suffering from eroding margins caused by the procurement platforms such as Amazon and German fashion specialist Zalando; after a while, it no longer makes sense to protect such a business model. To make things worse, omnichannel generally leads to added complexity and costs.

SCALE

"Nothing scales as well as a software business, and nothing creates a moat for that business more effectively than network effects."

– Tren Griffin, Two Powerful Mental Models (2016)

Transformational Products need to scale if they can hope to produce significant revenue and profit growth. It's one thing to find a business model that works, it's something completely different to make it scale. Transformational Products have the big advantage that they scale rather easily and that results can be tested and proven with comparatively small investments of time and money.

Economies of scale are the basis of modern mass production, and as such are well understood. Their theoretical underpinnings reach all the way back to Scottish economist Adam Smith and state that it is possible to achieve a perfect balance where overhead costs are spread over as many units of production as possible, beyond which point the marginal costs of producing additional units become minimal. In the case of physical products, in general,

marginal costs increase if this tipping point is passed – unlike digital products, where marginal costs eventually fall to zero. This not only allows to scale much faster but also means they can scale further. Addressable markets far exceed those for physical products.

One way scalability effects markets is that it often leads to the formation of natural monopolies. Classical example are public utilities (highways, tele-phones, postal services, energy and water supplies). Building these networks involves huge fixed costs, but the operating costs are comparatively low. The same goes for Transformational Products: it costs a lot to develop the software, but the marginal cost of running it in the cloud are negligible. In the case of networks, there are the network effects that correspond to the economy of scale benefits in physical products.

Transformational Products produce positive scaling effects both for the provider and the consumer. For the provider, this means sinking operating costs, whereas the customer profits mainly through scaling effects. The latter effects are often described as positive feedback. Metcalfe's Law states that the value of a telecommunications network is proportional to the square of the number of connected users. This axiom was long considered a mere rule of thumb but research using empirical data from Facebook and Tencent's WeChat have proven its validity.

Positive feedback strengthens the strengths and weakens the weak-nesses of a product. In extreme cases where the network effect is strong enough, this can lead to a winner-takes-all situation and an effective monopoly where the dominant player or a single technology takes over the entire market. GAFA in their respective core markets are good examples of this. Even Apple, despite its status as a hybrid company involved in both hardware and software, has managed to snatch up more than 100 percent of all profits made in the worldwide smartphone business in certain given periods, leading almost all other competitors to post negative results.

Only products that scale sufficiently can hope to deliver significant turnover and profits. Legacy business models need to scale the digital service layer if they want to stay alive. Once scale is achieved, additional partners can be added to the value chain via APIs, turning the Transformational Product into a platform to bring third-party supply and demand together. When this happens, data become the decisive factor, creating relevance for the user whose choice of the content, goods or services being displayed on the platform is pre-filtered and personalized.

APIs

"A 'platform' is a system that can be programmed and therefore customized by outside developers – users – and in that way, adapted to countless needs and niches that the platform's original developers could not have possibly contemplated, much less had time to accommodate."

– Marc Andreessen, The three kinds of platforms
you meet on the Internet (2007)

Conspiring to create value (co-creation) doesn't stop with customers and companies. Transformational Products should allow for customization by third-party developers and thereby adapt to countless needs and niches that lie outside the scope of their original creators. By covering a constantly growing number of use cases and connecting to other platforms the product creates even more value for its users. Expandability is the goal, and in Transformational Products, this is the job of APIs.

APIs, regardless of how they are implemented technically, are simply interfaces. In communication theory, they are defined as the point of interaction between systems. Transformational Products are not only a bundle of services, but of systems, too – and it's not just technical systems. Organizations also have APIs – and both intersect. This goes especially for large enterprises' APIs that create workflows over their entire range of resources and are aimed at harmonizing access to these organizational resources, which consist mainly of data or services.

Well-constructed APIs aren't just important for Transformational Products; APIs for organizations also require clean design. According to Gall's Law, all complex systems that work evolved from simpler systems that worked. Complex systems that are built from scratch seldom work and can hardly be repaired. Better, therefore, to start with something simple. That's why it's so hard to create Transformational Products with good APIs within the confines of legacy systems developed by pre-existing organizations. These attempts often fail due to seemingly banal services such as central login for all the company's digital products. APIs can be the greatest single challenge facing an enterprise on the way to digitization.

Polymath Herbert A. Simon, a Nobel Prize winner who was among the earliest to analyze the architecture of complexity, postulated that complex systems need to be simplified by separating them into smaller building blocks, each with a clearly defined interface to enable interaction. Focusing on APIs is therefore important not only for organizational structures but for technological infrastructures, as well. This suggests it is important to get the API right first and let all the rest follow.

There are internal and external factors that influence the development of APIs that need to be considered.

①   **Private APIs.** Private means integrating use within the confines of an organization. Various business units sharing a domain can use these interfaces to access the infrastructure of a product. Here, again, Amazon provides a perfect example. It was decided very early on that every new internal service should be represented by its own API. Over time, Bezos succeeded in creating a very robust internal platform infrastructure. Each business unit communicates with the others through this platform. Thus, Amazon represents a tight network of independent business units that are constantly in touch with one another and that collaborate closely, aided by a continuous flow of data. These APIs were so powerful they later enabled the company to become the world's leading cloud service provider as the backbone of AWS.

②   **Partner APIs.** The administered partner space refers to APIs that are open to third-party users, but only from partners that are closely supervised.

③   **Public APIs.** The openly accessible space is serviced through public APIs that anyone can use. Often, these APIs are used to provide self-service functions, but also to give third-party developers a means to interact with the platform (plug-and-play).

The obvious difference here is the degree of accessibility provided. Regardless of the use case there are five typical application scenarios in which APIs can operate as the gateway to platforms:

- Activation of mobile channels and IoT
- Creation of ecosystems
- Improved diffusion
- Developing new business models
- Forcing innovation

In every case, the API works as an invisible enabler creating new value for the customer.

DATA

"Data Love" was the motto of the German NEXT Conference 2011 which gave the old debate about data a new spin. The discussion had focused almost exclusively on privacy issues, and for a data protectionist, the design doctrine of choice is always less data, not more. But developing Transformational Products calls for a completely different concept, namely grabbing all the data you can get. The data isn't explicitly provided by the user but gathered implicitly through co-creation caused by the user's interaction with the product itself.

Data sets are the key to every successful digital product. They are the foundation upon which individualized services are built, enabling them to provide relevant benefits regardless of time and context. The depth and width, or capacity, of data form the vertical and horizontal reference lines in a system of coordinates describing a given data strategy. The further to the upper right-hand side of the diagram, the easier it will be to develop products that align seamlessly with the user's daily activities.

Fig. 8 shows examples of the location of data strategies in a few leading providers of digital products. Google Maps is a good example of how the vertical and horizontal importance of data impact the product's use value.

Vertically, Google Maps draws data from its linked user accounts. These user profiles are put to use in every conceivable device format (desktops, tablets, smartphones). Thanks to their geo-tracking function, smartphones are especially useful for collecting enormous amounts of personal data. As a result, Maps is exceptionally good at creating a personalized user experience because every user can see their very own map reflecting their day to day life; their places of residence and work, favorite restaurants, frequently used routes, and much more. All these details are displayed on their digital maps; less pertinent items of personal interest are held in the background but available in an instant.

Horizontally, Maps can combine usage patterns from hundreds of millions of users. In Europe and North America, Maps can boast a depth of data that not only enable them to create personalized maps, but also transform the service into an interactive platform. Google knows which way you like to drive or ride to work but it can also compute the best connections and routes for every available form of transport. To this end, the company collects real-time tracking data from other users who happen to be travelling in the vicinity, thus providing up-to-the-minute recommendations and timetables. Google can even provide information about how many people can be expected to view a named attraction or stand in line at a store or museum at certain times of day, all with an astonishing degree of accuracy.

Google Maps is also a good place to watch co-creation in action. Users voluntarily link the map service to their personal accounts and the tracking function of their smartphones to be able to use the service more conveniently, as well as more effectively. By doing so, they are supplying Google with valuable data points with which they can improve the quality of their experience – giving the service a virtually unbeatable edge over any competitor.

Gathering all the data you can is also important for another reason. In the age of artificial intelligence, data will be the next major battlefield. AI systems today are capable of self-learning, simply by analyzing large collections of data using staggered, feedback-enabled neural networks. All the big digital players – Google, Apple, Facebook, and Amazon – as well as the major technology companies like Microsoft and IBM are pouring billions of dollars into developing the algorithms of → machine learning – most of them destined

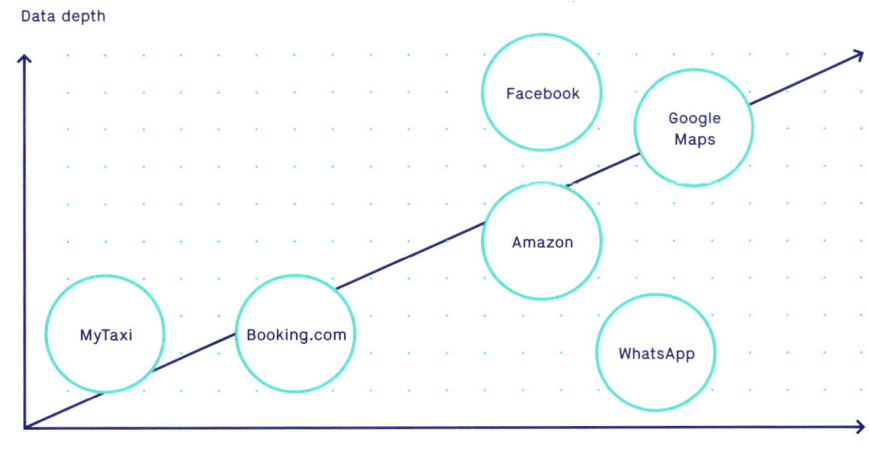

Fig. 8: Data depth vs. data width

to eventually become open source. This is because these companies want to draw in as much outside developer expertise as possible and lock them into their own particular technology, thus improving their position in the market for talent, which is expensive as well as increasingly hard to find. In addition, algorithms need huge amounts of data in order to achieve acceptable results. Here, again, the GAFA companies enjoy a comfortable lead. By utilizing these hoards of data they are able to make their products even more attractive to users. One day soon, Google Maps will be able to whistle up a cab for us without asking us first, just because it knows we have probably forgotten an important meeting and will have to hurry if we want to be on time.

Google has never asked us for explicit permission to personalize their services like this. Instead, personalization has crept up on us through a crafty combination of individual steps, each one seemingly plausible and potentially beneficiary. Personal user accounts and non-stop tracking of our movements aren't the result of any blank check given them by the users, but instead are the result of a series of convenience nuggets users are eager to profit from. Any enterprise seeking to cut out its share of these markets will need to provide at least an equivalent user experience; if not, the odds are their products will fail, simply because it will not be the most convenient one for the user. In future, the company with the deepest and widest troves of data will be the one that prevails because it can provide the best product experience.

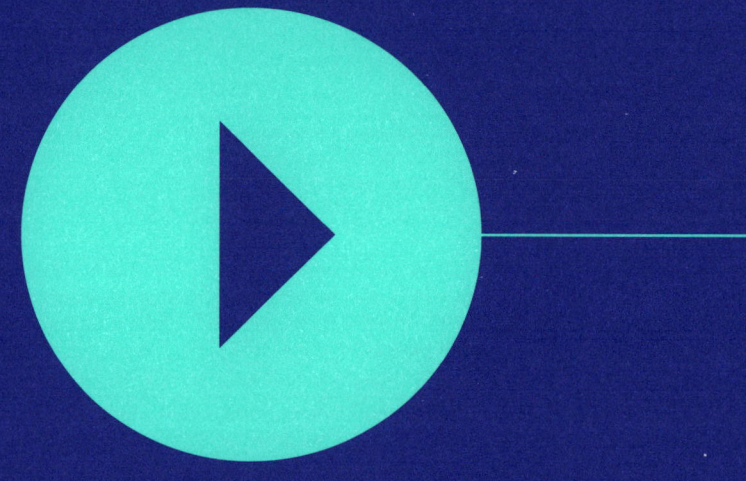

PLAYBOOK

Building Blocks

"Without a good process it's difficult to achieve a good result, but the resulting product itself is so much more important!"

– Michael Butlitsky, The World is a Product (2013)

How are successful digital products created? At SinnerSchrader this is a question we have been asking ourselves time and time again for more than two decades. We have learned a number of lessons in the meantime, but one stands above all others: there are no shortcuts. Product development always means hard work, and as with every innovation, there is no guarantee of success.

When attempting to create an innovative product, always start by trying to discover a new benefit for the customer. The danger here, however, is that the use of popular methods such as → Design Thinking – remember all those bright-colored sticky notes on the walls of conference rooms? – often lead to focusing too much on ideation. Taken to extremes, this can turn product development into a new-fashioned workshop format. And while it may be possible to address real customer needs at such a rap session, the results often have little or nothing to do with the company and its context, much less with scalability and value propositions. They remain irrelevant to the company's future.

Drawing on our long experience of working with hundreds of startups as well as Fortune 500 or DAX companies, the team at SinnerSchrader has put together a playbook for designing Transformational Products that focuses on methods and how best to keep one's eye on the ball – the product. In it, we strike a balance between diffusion, experience and co-creation and use the Product Field method, a comprehensive model and tested toolbox for Product Thinking, to make certain that a company's legacy and strategic context within its specific industry are not neglected over the course of product development.

The building blocks of our playbook are the following:

① **Product Team.** This is the crucial success factor. It involves not only bringing together talents from many different fields of expertise, but above all infusing them as a team with a common vision of the product.

② **Product Creating.** This is where the Transformational Product is identified, where it evolves and finally becomes the pilot to be tested in the market, or with a sufficiently large audience. In this phase, → end-to-end responsibility lies with the product team that also determines the methods to be employed, usually Product Field and Product Toolbox.

— **Product Staging.** The EXPERIENCE LOOP provides the structure for step-by-step product development and is iterated repeatedly over the course of the product creation process.

— **Product Field.** Product creation is never a greenfield process, but occurs within a given context of resources, markets and innovation drivers. The Product Field method uncovers the forces at work and validates the results delivered by the product team.

— **Product Toolbox.** Here we introduce a set of helpful tools for research, implementation, and optimization.

③ **Product Factory.** To be successful, Transformational Products need to correlate to the success of the enterprise as a whole, feeding back into the organization itself. Gearing up for industrial-scale production follows its own criteria that must be included in the organizational design.

Fig. 9 on the next double page shows how these building blocks fit together.

 ① PRODUCT TEAM → ② → ③ PRODUCT FACTORY

Product
Management

Company
Transformation

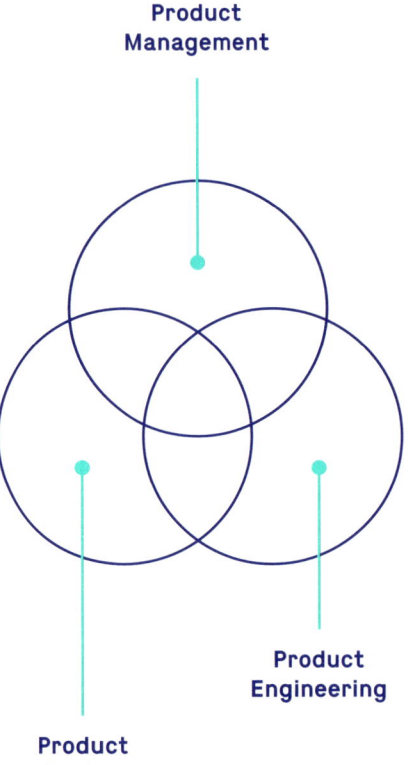

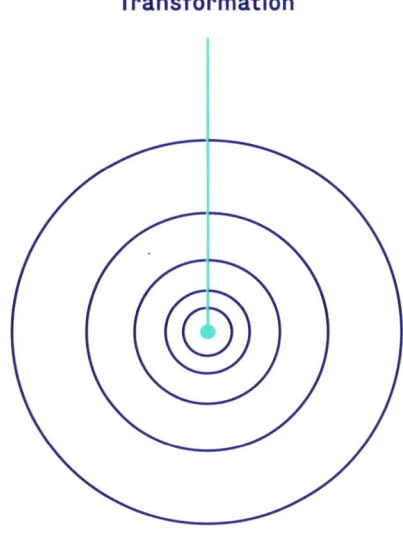

Product
Engineering

Product
Design

Fig. 9: Playbook at a glance

② PRODUCT CREATING

PRODUCT FIELD

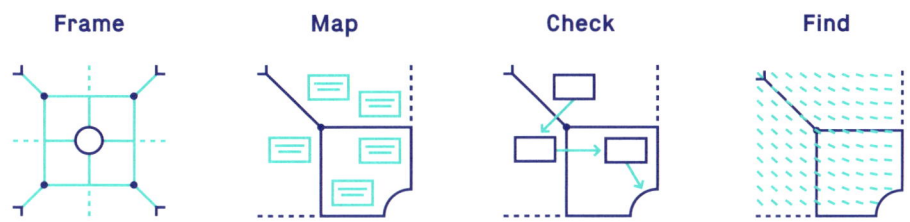

| Frame | Map | Check | Find |

PRODUCT STAGING

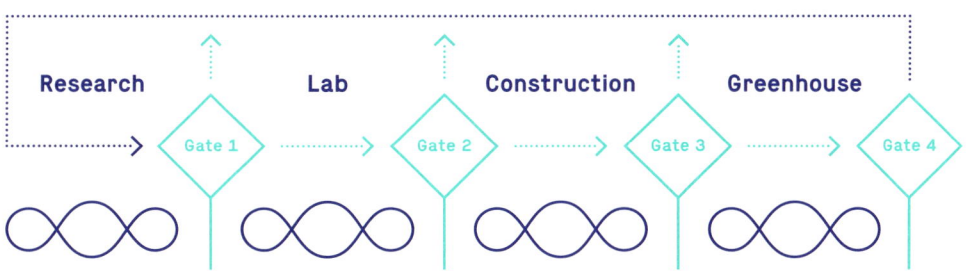

Research — Gate 1 — Lab — Gate 2 — Construction — Gate 3 — Greenhouse — Gate 4

PRODUCT TOOLBOX

- Design Thinking
- Service Design
- Prototyping
- Design Sprint
- Agile Development
- Lean
- Testing
- Lean Analytics

Product Team

The product team plays the most critical part in developing Transformational Products. Without the right cast, you can essentially forget the whole thing. The team needs to consist mainly of product people. These are individuals that think in terms of products, not processes. Naturally, they are masters of the methods and tools necessary for product development, and we will discuss the main ones in the Product Toolbox chapter. But they also know that, in real life, it is necessary to cut processes some slack. Process rules are not set in stone and shouldn't be taken too literally: they must allow the product to take precedence. After all, it's not about processes, it's about products and their value for customers. The product team should therefore be free to choose which methods from the product toolbox they want to employ. They're just tools, after all.

Thinking in products is completely unlike thinking in projects. Project thinking focuses on (project) processes, timing and resources. Management becomes the key discipline because projects have a beginning and an end. Successful products, on the other hand, outlive their project teams. At the end of this chapter we will explore how this transfers to the organization as a whole.

The signature of a good product team is its interdisciplinary composition. These we refer to as → full-stack product teams. We have borrowed the term from Chris Dixon, an associate at venture capitalist firm Andreessen Horowitz, who described a new kind of founder team:

"Full-stack founders care about every aspect of their product/service, so they need to get good at many different things besides software – hardware, design, consumer marketing, supply chain management, sales, partnerships, regulation, etc."

Every member of the product team, whether hailing from management, design, or engineering, needs to be up to the challenges. This means they need to be

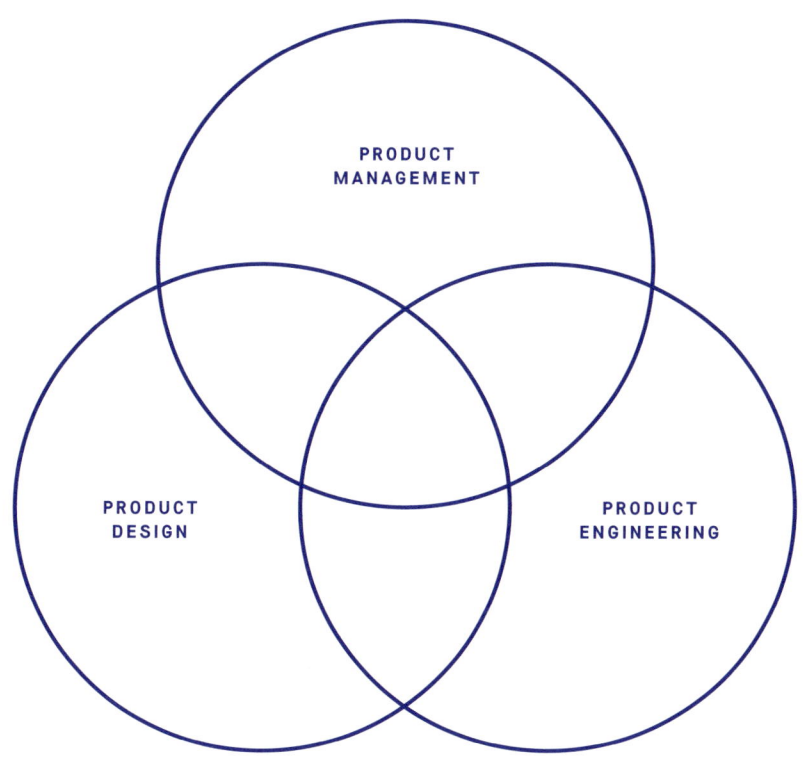

PRODUCT
MANAGEMENT

PRODUCT
DESIGN

PRODUCT
ENGINEERING

PRODUCT MANAGEMENT

- Business strategists
- Data analysts
- Product manager

PRODUCT DESIGN

- Brand & identity
- Business & service
- Process & architecture
- Interaction & interface

PRODUCT ENGINEERING

- User interface
- Mobile
- API design
- Cloud integration

Fig. 10: Product Team

experts in one or many areas but also bring general knowledge about a wide range of adjacent topics to the table; areas in which they are not necessarily experts, but have a good understanding which enables them to achieve an overarching sense of the interdependencies and to work the various interfaces to create a unified vision of the desired outcome and how best to reach it. This is the only way to successfully prioritize the various artefacts and contributions that need to go into it.

It's important to avoid obsessing too much on ideation and design, but instead to focus on developing a robust product. This means finding the right balance between implementation, diffusion, scale and business modelling, among others. The rise of the big platforms has led to greatly increased customer expectations, and if you don't find the right mix between product management, product design and product engineering from day one, it will be hard or even impossible to correct it later. It makes sense, therefore, to give the product team end-to-end responsibility for everything from research to the demonstration of product market maturity, and not just to bring them in during the exploratory phase of product development.

In our experience, product teams work best in studio environments similar to those preferred by startups and creative agencies. In isolation, teams can work without the disturbing hindrances of in-house politics and the distractions of everyday business. Being able to exchange and interact with parallel product teams within the environs of a large studio can be extremely valuable. We will now consider the various roles that play a crucial part within the product team.

PRODUCT MANAGEMENT

"I like my product managers to focus on the most miserable thing people have to deal with everyday. If you can solve that problem, that actually changes behavior, and that can lead to the truly big product wins."

– Jeff Bonforte, in: Marty Cagan, Inspired (2008)

Among the first to practice product management for marketing consumer goods were Procter & Gamble. Neil H. McElroy was the author of the famous *Brand Man Memo* (1931), in which he argued that each brand within P&G should be marketed as if it were a separate business – a new idea at the time, and one which was destined to become a key element of modern marketing in the FMCG industry. The Brand Man, as McElroy became known, later consulted two young hopefuls in the IT industry Bill Hewlett and David Packard, the founders of one of the most successful computer companies in the world.

Decisions were taken as close to the customer as possible at Hewlett-Packard (HP) and product managers were the voice of the customer within the company. HP became the most customer-centered business within the computer industry. Following World War II, Toyota combined the concepts of just-in-time production and continuous improvement within its Kaizen, literally "change for better", philosophy and HP was one of the first big companies in America to take the idea on board. This new understanding of product management and production gradually spread from HP to other young IT companies, and it's probably one of Silicon Valley's lesser-known but critically important success factors.

In consumer goods, product managers are typically attached to marketing. Within the classical marketing mix with its Four Ps (product, place, price, promotion), they tend to focus on the last three. As a rule, product development is completely separated from this, but in the digital world things don't work that way and product managers need to focus on the product itself. The objective is not only to understand the consumers and their needs in terms of how to sell it to them, but also to put the customer at the very center of the development process. If at all, promotion and branding are only reluctantly included, in the belief that only inferior products are deemed in need of costly advertising.

The difference between consumer brands and technology companies in terms of product management is clear-cut. In those industries that are beginning to reinvent themselves as software companies, like automotive, financial services, telecommunications, retail, and tourism, the definition of product management often hinges on the professional biographies of top leadership.

Our own understanding of product management is strongly influenced by Silicon Valley's approach to the subject. Its role can be described this way: Product management is responsible for discovering a product that is valuable, beneficial, and → feasible. Product discovery is the result of collaboration between product management, design, and engineering. Once a real user problem is identified, and thereby a new opportunity, product management has four fundamental responsibilities, according to our playbook:

① **Concretize** the product idea by defining the necessary steps to take in its staging.
② **Validate and adjust** the product according to the Product Field.
③ **Determine** what needs to be built.
④ **Align stakeholders** by preparing the necessary artifacts (→ stakeholder alignment).

The third point in this list is the tricky one. It requires lots of discipline, keeping one's eye firmly on the ball – specifically focusing on the user and their problem. Loss of focus, for instance through feature overkill, not only bogs the process down, but can also put the kibosh on the product's later market success.

Product managers need to be passionate about products. They need to live, eat and breathe products. They must love products and respect them, no matter where they're from or to which category they belong. In fact, a good product manager doesn't even have to be part of the industry to which it belongs and for which it is being developed. What they do need is to be on the same wavelength as the product's user and to take pride in the company that makes it. What this means is that a good product manager must be able to immerse himself completely in the customer's mindset and to think solutions all the way through. Simply studying the customer is not enough. Steve Jobs famously made this point when he said, "customers don't know what they want until you show it to them."

PRODUCT DESIGN

Good product design, especially when it comes to digital products, is distinguished less by formal aesthetic design and more by an outstanding → user experience design (UXD), which becomes an important quality in itself. Experience over recent years points to four roles that play big parts in the designing of successful digital products:

- Service designer (SD)
- Interaction designer (ID)
- Visual designer (VD)
- UI developer (UID)

This viewpoint is only of limited use considering the challenges facing today's generation of product designers with regard to multi-device/multi-channel capabilities, information-gathering bots, IoT, voice interfaces, invisible interfaces, AI, data, and so on, as well as the need for → agile, cross-functional product design teams. The necessary high degree of overlap in terms of competencies, skill sets, methodologies and tools within a modern multimodal product design team call for more-open role descriptions. These tend to follow the idea of a networked economy where maximum value is extracted from a complex mix of assets, requirements and individual talents.

Experience designers (XD) are crucial to creating successful digital products. These individuals are usually multi-talented, and their résumés are marked by an openness for new developments. In fact, all members of the product design team should see themselves as experience designers sharing a common goal of creating an outstanding user experience design (UXD) instead of withdrawing into their respective silos of expertise. There are always opportunities for close cooperation with product engineering for team members with very different areas of competence, depending on the product's stage of development. Creating cross-functional teams is a good way to make this happen.

Digital product design involves these competencies, among others:

- Brand & identity
- Business & service
- Process & architecture
- Interaction & interface

The design process shifts over time from the abstract (brand & identity) to the tangible (interaction & interface). This is usually a stepping process characterized by short release cycles and immediate feedback in which data from earlier cycles flows back into the design process through an ongoing basis. Experience designers with a command of the full set of competencies, abilities, methodologies and tools are relatively rare today. This will change as time passes.

We will now discuss these fields of expertise in more detail.

Brand & identity

Developing brand identity for Transformational Products is very different from the process followed for traditional products – in fact, it turns the process on its head. Digital products do not derive from a given brand identity which is associated with certain values, target groups, styles and narratives.

Transformational Products are created by focusing closely on solving a very specific user problem and developing a radically new and improved user experience. They do not distinguish themselves through brand identity, but through offering as many users as possible a better way to get something done. Achieving network effects and better use value are key if a product is to excel.

The design system thus created involves graphics, motion, sound, interaction, tonality, and space, and aims at delivering maximum benefit to the customer. Therefore, things like utility, usability, desirability, constant analysis of user data and the resulting improvements to product design are important. Factors like an established corporate design system or brand identity awareness are usually no big help.

Google's string of successful Transformational Products serves to illustrate this point. At first, the company chose to focus on purely digital products such as Search, Analytics, Mail, Maps, Docs, Sheets, and Photos. The use value of each was constantly improved, but only later did the company develop Material Design. This unique design language spans the entire portfolio and provides an even better user experience for both its platform and its products, which reflects to the Google brand itself and reinforces its brand identity.

Over the years, this has led to a wide range of style guides, pattern libraries and handbooks based on successful product performance instead of vice versa. It is this ability to think strategically, along with superb craftsmanship, that marks the successful experience designer in the field of brand & identity.

Business & service

What does the customer need, what are the obstacles faced and what are the alternatives? What does the market offer today and who are the major competitors? What is the operational environment? Which key personalities can be identified? What are the most important use cases and what kind of operational framework will prevail? What is the user's problem to be solved, preferably better than ever before? Which business models devolve from this? What are we talking about here, anyway? Business & service is all about design strategy, expressed through product models and narratives. This cannot happen without managing change throughout the existing organization.

Process & architecture

The next step towards creating an actual product involves moving from strategy to tactics. What does this mean? Do processes and an architecture need to be developed for it, and if so, which ones? The process & architecture competency calls for clever and calculated product design, including in-depth analysis of its potential as well as designing the required business processes and the functional architecture to go with it. Experience designers whose strength lies in process & architecture are the ones that will write the necessary business specifications and describe their functionality (interface architecture, user journeys, use cases).

Interaction & interface

The last link in this chain that leads to successful product design involves creating the right kind of interface between the product and the customer. How will users interact with the product to achieve the desired result? For this, we need use cases, wireframes, models, mockups and prototypes. This competency involves all aspects of the product's perception across various different applicable categories to bring it to life. Naturally, this involves quantitative and qualitative user tests, as well.

PRODUCT ENGINEERING

Development processes are traditionally the result of past project work, which explains their decades of predominance in software development where the apportioning of time, effort and budget are determined by them. The world of software development classically consists of separate units for frontend, backend, database, quality assurance and operations. This approach makes no sense at all in the context of product engineering as we see it, nor does the recent trend towards end-to-end or full-stack development provide the answer, at least not in isolation. Besides the necessary technical skills, team members must demonstrate a wide range of qualifications as described above. This is the kind of profile needed at the intersection of product management and design to produce the best concepts that enable the product engineering team not only to make the product happen, but also to ensure a professional implementation. This involves more than asking the simple question "What is technically feasible?"

Despite its new operational framework, it goes without saying that the product engineering team still needs plenty of know-how in the various classical technical disciplines. Our experience shows that these must include the following competencies:

User interface

Even though the borders between frontend and backend development are becoming blurred, or even disappearing altogether, the ability to program a good graphical user interface for interaction between humans and machines, using HTML, CSS, or JavaScript for instance, remains important and is critical for a good user experience in terms of product performance and compatibility. Mastery in this area calls for a certain elegance (don't code things twice), as well as a proper understanding of code maintenance.

Ideally, the user interface (UI) engineer and the experience designer responsible for interaction & interface will form a close unit at an early stage of the process. This allows interface solutions to be designed, coded and tested at an early stage, either in the browser or the app. Sophisticated Atomic Design methodology and so-called → living style guides help document styles and patterns, keep designers and developers in sync, and greatly help to organize and distill complex interfaces. Usability and performance problems, in particular, can be identified at an early stage and various alternatives can be coded quickly and easily using the living style guide.

In recent years, attention within the developer community has been increasingly focused on the user interface and its implementation, adopting a user-first mentality. It would, however, be wrong to suppose that technology has therefore become less relevant but it does need to be brought in line with the user. Facebook's ReactJS, a JavaScript library for building user interfaces, and similar tools are becoming increasingly popular with web developers who are becoming more prone to following a common-sense approach involving

tools like ReactJS, Redux, JSX and NodeJS for managing both data and UI states in applications.

Ultimately, we believe it is important for the technology employed to fit the experience, capabilities and personal preferences of the development team. Neglecting to consult your product engineering team before introducing a new tool is never a good idea.

Mobile

A (native) mobile developer, in our opinion, is simply a UI developer with a special set of skills.

API design

The job of an engineer in designing APIs goes far beyond simply building an interface between systems or UI development endpoints. There is more to integrating existing services or creating new ones than just getting the technical part right. Good API design needs to reflect the functional and commercial vision of the product. The aim of good architectural design is to smooth the workflow during the development process and increase efficiency. This still calls for a high degree of technical expertise, but that isn't all.

Cloud integration

Software developers often refer to something called "native cloud architecture". What they mean is the focused interplay between microservices, cloud/DevOps (development operations), and continuous delivery in order to create scalable applications. One of the best ways to do this is the Twelve Factor method involving a list of principles, each explaining the ideal way to handle a subset of an application, which was developed in 2011 by Heroku's co-founder Adam Wiggins.

Product Creating

PRODUCT STAGING

The product should always take center stage and product staging, unsurprisingly, is central to this playbook and well-deserves our attention. We distinguish four separate → stages, and product teams need to emphasize all of them, both in sequence and in combination.

- ◇₁ **Research**
- ◇₂ **Lab**
- ◇₃ **Construction**
- ◇₄ **Greenhouse**

Dead-ends and loops are integral and laudable parts of the development process. Contrary to the user journey that follows the **EXPERIENCE LOOP**, it is necessary to go through these stages in reverse order, namely from **SERVICE CO-CREATION** and **SERVICE EXPERIENCE** to **SERVICE DIFFUSION**.

The product team is best advised to validate their results by using the Product Field, as well as other tools from the product toolbox. The **EXPERIENCE LOOP** should be repeated several times until the → gates defined by the product team have been passed (see fig. 11). We will suggest a number of criteria for defining the four stage-gates later.

Research

In our experience, the research stage may require some experimentation before a consistent **EXPERIENCE LOOP** is found. There is also a convenient way to retrace one's steps from the **SERVICE DIFFUSION** stage to the two initial stages (**SERVICE CO-CREATION** and **SERVICE EXPERIENCE**) if this becomes necessary. The best business model and the most perfect customer value miss their mark if no lever for successful product diffusion can be found.

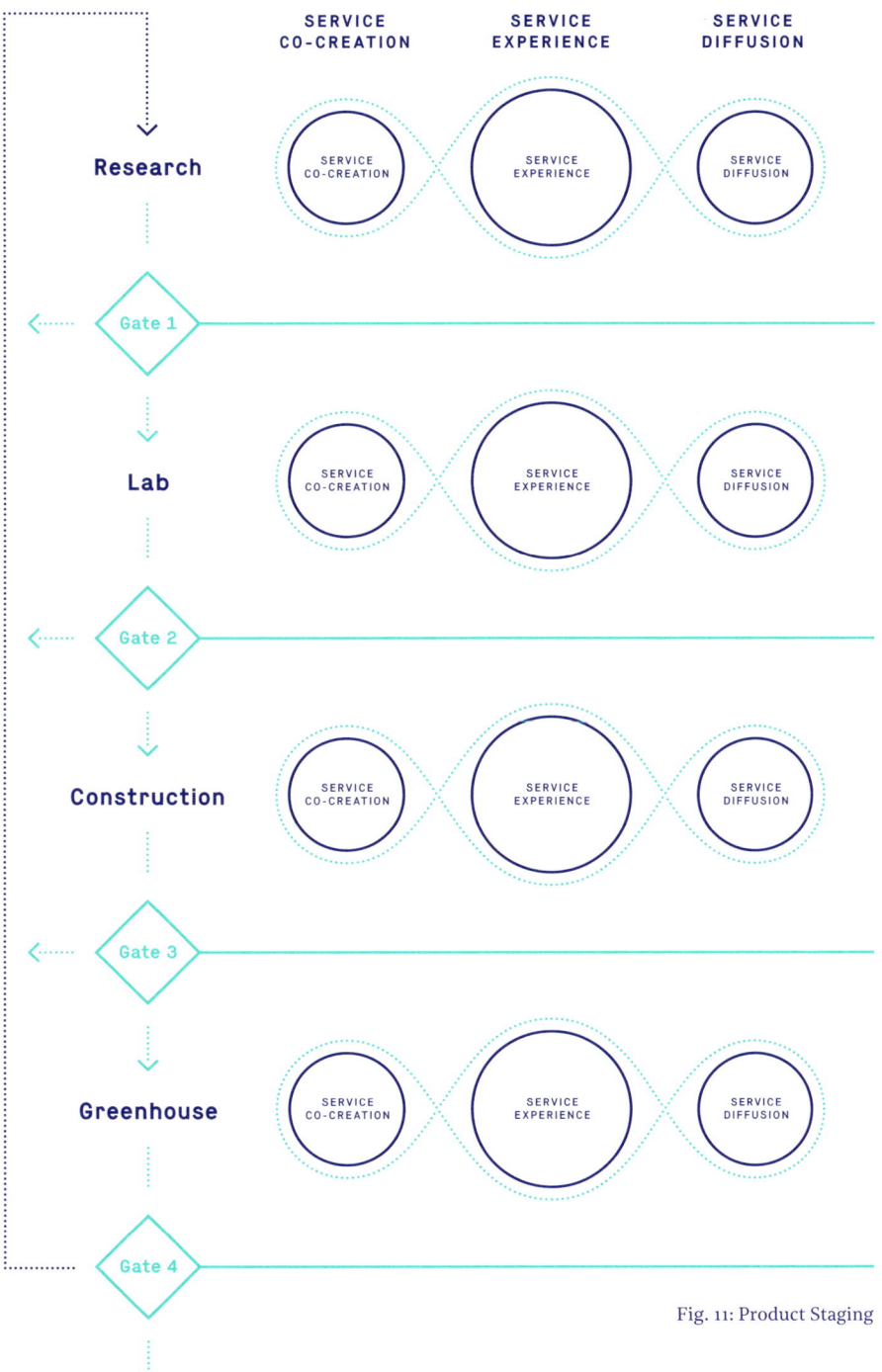

Fig. 11: Product Staging

∞ SERVICE CO-CREATION

For every process, there is a starting point that anchors the iterative cycles to follow. In our playbook this is **SERVICE CO-CREATION**, which starts with the discovery of a product's use value. If at this stage there are still no bright ideas to serve as stepping stones for the further development process, that's okay. There are numerous methods that can assist in ideation, and we will introduce two of these in the chapter dealing with the product toolbox. Crucially, the discovered use value not only needs to be novel, it should also be proprietary in the sense that it can only be fully achieved in conjunction with the specific resources of the company. The following graphic illustrates this principle:

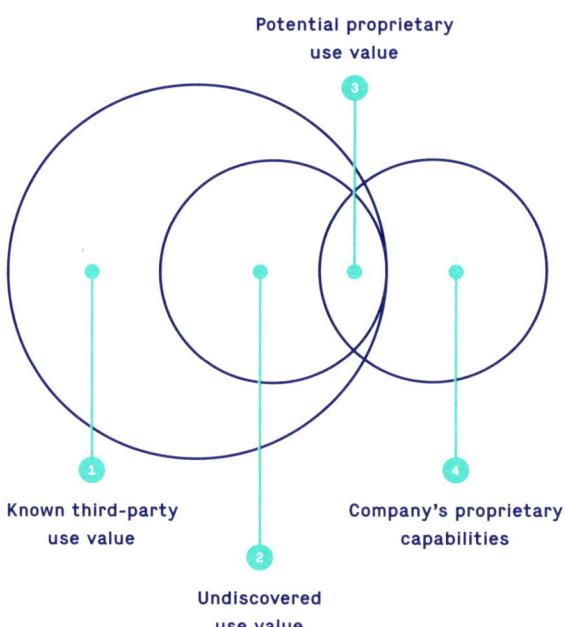

Fig. 12: Discovering use value

① Stiff competition: amount of use value already discovered
② Potential competition: amount of potential use value regardless of company resources
③ Unique market position: amount of potential use value that can only be achieved in conjunction with company resources

We start with the product definition as the outcome of our discovery process through which we found the product's proprietary use value which only
our company can deliver. This is the point that will provide differentiation
throughout the product staging process. Startups work differently, by the way,
but we are not concerned here with developing Transformational Products
from scratch as they do.

In the next step, we seek to work out the key success factors that form
the code of a Transformational Product. A good way to do this is by answering
the basic questions for every product definition: Why, What, and How?

- [ ] **Why** does the product create value?
  - [ ] **What** is the new use value?
    - [ ] **What** specific company resources are necessary?
      - [ ] **How** will we earn money? (⟶ business model)?
      - [ ] **How** can we scale? (⟶ scale)?
  - [ ] **What** do the external services look like that we need to integrate?
    - [ ] **How** do these services change the value creation model?
      - [ ] **How** can we generate valuable data? (⟶ data)?
      - [ ] **What** technology do we need for integration? (⟶ APIs)?
- [x] All these questions need to be answered.

✕✕ SERVICE EXPERIENCE

In the research stage, the whole product team should share a common vision of what kind of product experience is to be delivered, while making sure they avoid falling into the trap of premature prototyping. Essentially, everybody should be working with pencil and paper at this stage. The following questions need to be addressed:

☐ **Why** does the product change user habits?

  ☐ **How** can we reduce the threshold for early usage?

   ☐ **How** would a novice start using the product?

    ☐ **How** do users interact with the product? (⟶ user interface)?

    ☐ **How** does the product trigger a positive reaction? (⟶ user experience)?

  ☐ **How** can we market the product most efficiently?

   ☐ **How** is the user rewarded through positive feedback?

    ☐ **How** do we achieve ⟶ mental lock-in?

    ☐ **How** do we achieve ⟶ functional lock-in?

☑ All these questions need to be answered.

✕ SERVICE DIFFUSION

Transformational Products gain hold of markets mainly through use. Classical marketing only serves as a catalyst. However, this process doesn't happen spontaneously, it requires powerful levers that need to be developed early in the research stage. We cannot stress enough that **SERVICE DIFFUSION** demands that important marketing functions, for which classical products employ external marketing strategies, must be woven into the fabric of a Transformational Product – the product must be self-marketing.

First attempts at **SERVICE DIFFUSION** will almost certainly lead to a dead end: The values identified in the **SERVICE CO-CREATION** process contain no mechanisms that lend themselves to successful market penetration, and recognizing this fact will save a lot of time and resources. Instead, the product team should return to the research stage and start again at the level of **SERVICE CO-CREATION**.

- ☐ **How** does the product alter user expectations?

  - ☐ **How** can we make the product easier to use?

    - ☐ **What** are the triggers?

      - ☐ **What** kind of user experience will differentiate the product from others? (⟶ radical value proposition)?

      - ☐ **How** can we achieve superior usability? (⟶ casualness)?

  - ☐ **How** can we guarantee marketing efficiency?

    - ☐ **What** will change user habits?

      - ☐ **How** do we unlock new dimensions of use value? (⟶ 10x value)?

      - ☐ **How** can we make marketing part of the product? (⟶ built-in marketing)?

- ☑ All these questions need to be answered.

Lab

This playbook concentrates on the product with the goal of ensuring visibility and guidance on how to make it popular. Unfortunately, the product definition arrived at with pencil and paper during the research stage, along with the (textual) answers to the questions asked above are only effective within the product team itself.

Prototyping is essentially a communications tool and can help foster a common understanding of the product among the members of the team as well as between the different stakeholders within the company. Unlike the usual mind-numbing visual presentation decks and paper-based requirements lists that nobody reads anyway, a prototype is something everyone can touch and feel. It can help assuage misunderstandings at different levels and gives an early glimpse of the user experience, and provides a benchmark for the product team from which they can validate progress, demonstrate value to the stakeholders, and generate useful feedback.

Prototyping is always an iterative process and provides increasing degrees of → fidelity, which describes a prototype's degree of closeness to the final product. In general, we distinguish between two levels or degrees of fidelity:

1. **Low fidelity (lo-fi).** Here it is already good to abandon intermediary steps like paper or slide shows and move directly to digital prototyping, especially in developing for gadgets like a smartphone. By spending lots of time in front of the computer screen the product team can start to establish a feeling for the subsequent user experience. Tools are vital at this stage to achieve steady improvement. There are numerous tried and tested prototyping tools available to assist in parallel (or lo-fi) prototyping, a method that requires hardly any special prior knowledge. However, lo-fi prototyping will always endure as a tool for internal team communications.

2. **High fidelity (hi-fi).** If realistic user feedback is what you need, then hi-fi prototyping is indispensable. User expectations of digital products

have become so demanding that feedback from lo-fi prototypes, which reflect an artificial laboratory situation, hardly allow for reliable evaluation of the product's eventual reception. Testing eventual user responses using hi-fi prototypes is a highly efficient way of defining parameters for the later stages of implementation. It will be necessary, however, to develop native prototypes for every target platform (web, iOS, Android, etc.).

◆ ③ Construction

The biggest challenge for every product team is the third stage, with its necessary focus on the more technical aspects, where the momentum generated in the research and lab phases needs to be maintained. The early development phases should have produced a solid foundation for the later stages of product implementation, but after all, the real product still needs to be built. Special attention should be devoted to the following considerations:

① **Team scaling.** Once the implementation stage is reached product teams and especially the engineering side tend to expand disproportionately. Special care must be taken to ensure that new members are well-integrated and made familiar with results achieved so far. Those joining up need to become fully committed to the common goal.

② **Methodology.** As increasing emphasis is placed on engineering there is a danger that focus may shift from the product itself to the methodology employed. However, it isn't just important to do things the right way (methodology), but even more to do the right things. If members of the team lose sight of the objective and start focusing instead on methods and processes, the entire team will soon be in trouble.

③ **Iteration.** During implementation fewer iterations may be seen than during the other stages of development. However, they do exist. While developing the interface, for instance, many tiny → tweaks will usually be made that in sum can hugely influence the way customers later experience the product. It is good to constantly question the status quo and demand suggestions for further improvements to the customer experience from all members of the team. But also when it comes to the integration of APIs and third-party systems, the generation and use of data as well as scaling and performance issues, there are often blockers in practice that cannot always be pushed aside. Impediments can seriously hamper the product's further development, and product management must demonstrate the courage to send the product back to the research and to the lab stages if necessary. Innovative products naturally create more implementation risks simply because they are, after all, → edge cases.

④ **Measurement, testing and optimization.** Once implementation is done and final testing is over, the product team will need to present the product to the target audience and the market at large. In anticipation of the necessity to deal with improvement suggestions, attention must be given during the implementation phase to a few non-functional requirements. These include the likelihood that the product will come under heavy scrutiny and preparations must be made for this, for instance by allowing for parallel testing (A/B and multivariate testing).

◆ Greenhouse

First contact between the product and real-life customers seldom goes as planned. Humans will often use a product in ways that the developers never anticipated and for which no tests have been made. Transformational Products are rarely ready to launch the day they leave the lab. They need to be tenderly nurtured by the product team, preferably in a safe enclosed environment.

We call this the greenhouse stage. This means that a Transformational Product is best launched to a focus group, hand-picked audience, or market. For Amazon in 1994, this meant North-American book buyers. The product team must then identity lever or set screws that will enable them to create a viable business model that will scale as needed (industrialization). This will initially be based on the business model developed during the research stage. Are those assumptions correct, and do the key performance indicators work as anticipated? This is hardly ever the case. As a general rule, optimization will require lots of work and will need incremental efforts to reach the stated goal. But even in those rare instances where assumptions prove correct, optimization can provide significant improvements.

Optimizing transformational digital products involves balancing interlaced considerations. Things like → cost per acquisition (CPA), → conversion rates (CR), → average revenue per user (ARPU) and → monthly active users (MAUs) are the usual indicators for a desired level of performance. Other drivers include price, competition, and many more elements that can vary according to field and business model. Getting all of them right usually calls for lots of experience and professional discipline – and sometimes the ability to recognize that there is no sense in flogging a dead horse, so it may be better to call it quits and start over.

Stage Gates

The product team needs to define stage gates (milestones) that must be reached before it is possible to proceed to the next level of development. It may be necessary to redefine these gates over the course of development by adding detail. Here is a very simple table listing some common generic stage gates:

Stage	Gate
Reseach	answer basic questions correctly
Lab	successful user tests using hi-fi prototypes
Construction	implement key requirements
Greenhouse	successful diffusion, growing retention

Product Field

Authors

Klaus-Peter Frahm

Michael Schieben

Wolfgang Wopperer-Beholz

We have learned by now that there are no good shortcuts to product innovation. Equally, there is no single silver bullet that will ensure success. If there was, then why do 80-90 percent of all startups, as well as attempts at innovation within established companies, fail sooner or later? Obviously, many of the best practices and process-frameworks for product innovation simply aren't reliable enough. Could it be that these methods are being applied too half-heartedly? Or are those involved just incompetent? We can't rule either of these answers out completely but our own experience and many case studies indicate there is another explanation.

An important reason one-size-fits-all practices and processes don't actually fit is because each product innovation needs to be seen within its own special context. Successful innovation is subject to the interdependencies and idiosyncrasies of the people involved, to organizational and market pres-

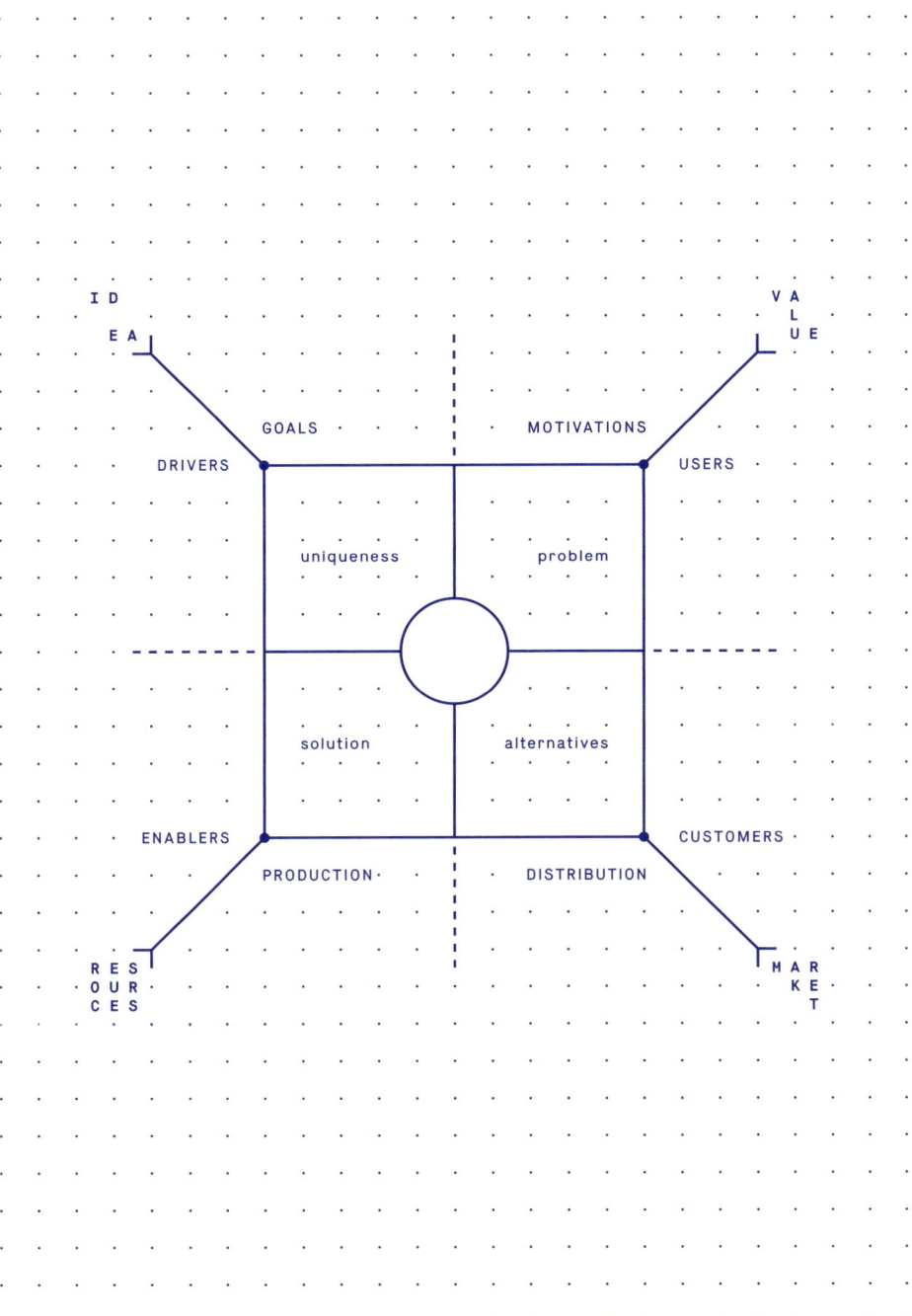

IDEA

VALUE

GOALS · · · · MOTIVATIONS

DRIVERS · · · USERS

uniqueness · · problem

solution · · alternatives

ENABLERS · · CUSTOMERS ·

PRODUCTION· · · DISTRIBUTION

RESOURCES

MARKET

Fig. 13: Product Field

sures, and a reliance on a whole raft of different technologies. Put another way, product innovation is a complex system that defies management by simple, cookie-cutter practices and processes.

It's questions like these that have led us to develop what we call the Product Field approach. How can we come to grips with complexity? If all these wonderful procedures for product innovation fail to function reliably, how should we proceed? Which tools and methodologies can we trust to make the interactive elements and forces within product innovation transparent and clear? How can we attune the thinking processes and the actions of those involved so as to unleash positive, dynamic innovation within a given system?

It seems obvious that the instruments used need to be made from a different mold than practical and processual recommendations based on a few more or less arbitrarily selected best-case instances. The tools needed to do a proper job must reinforce common, collaborative thinking. They must provide a vocabulary and a model that will lead to common understanding and help coordinate common action. They should promote analysis and evaluation within complex and dynamically changing systems. They must enable us to explore risks and potentials in a systematic fashion. And they should assist decision makers to find the right measures and methods and to determine when and how to employ them.

We call these tools cognitive media.

The Product Field is a cognitive medium. Since its inception in 2014 it has served as a comprehensive model and as a toolkit for effective Product Thinking. It is a huge help in dealing with challenges like these:

1. **Create a common frame of reference and shared understanding.** The Product Field offers a mental model, a visual structure and a shared vocabulary. We call this *frame*.

2. **Produce a big picture for all to work from.** The Product Field provides a → canvas on which to project available knowledge about product innovation and identify existing competencies. This we call *map*.

3 **Discover conceptual flaws, gaps, and inconsistencies.** The Product Field's inherent grammar makes it possible to validate an innovative product's coherence and consistency. This we refer to as a *check*.

4 **Disclose strengths and weaknesses.** The Product Field is good for discovering possible benefits as well as potential negative consequences within a given strategic, organizational or operative context and for defining action points to be taken. This we call *find*.

The following section will show how to use the Product Field and how the four steps – *frame, map, check, and find* – all fit together.

1 # FRAME —
Common frame of reference and shared understanding

To be successful as product creators we need to reach a common understanding of what product innovation is really about. The Product Field supplies a model to work from. It enables product teams and stakeholders to grasp the situation they find themselves in and to coordinate the decision-making process. In that sense, the Product Field is an instrument not only for analyzing and orientation, but also for conceiving and controlling product innovation.

Concept space and canvas

The Product Field provides a concept space with which to visualize the interrelations between agents, strategies, artefacts, and conditions. These can be projected onto a canvas and thus be made easier for people to grasp.

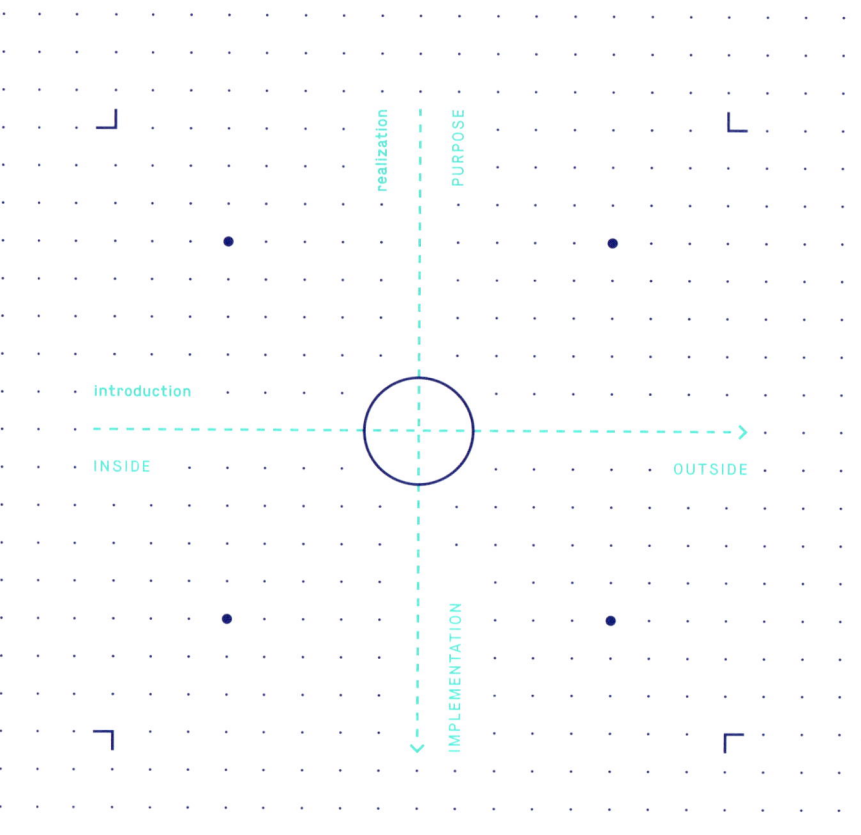

The space is structured by two orthogonal vectors.

INSIDE ⟶ OUTSIDE

Products are created inside a company but are sold to customers and for use outside that company. We call the vector from inside to outside *introduction*.

PURPOSE ⟶ IMPLEMENTATION

Products are created to fulfill an intent and purpose on behalf of its stake-holders. This vector leads from purpose to the product's implementation. We call it *realization*.

The extension of these two vectors create the axis of a coordinate system that rises from the midpoint between inside and outside, and from between the product's purpose and its realization. Thus, every possible permutation of product innovation is covered, and this is the Product Field's visual framework.

Aspects and fields

The Product Field describes every condition and force effecting product innovation. In order to cover and explore these in a systematic way, they must be arranged according to their cause and effect. Thus, twelve aspects of product innovation are identified.

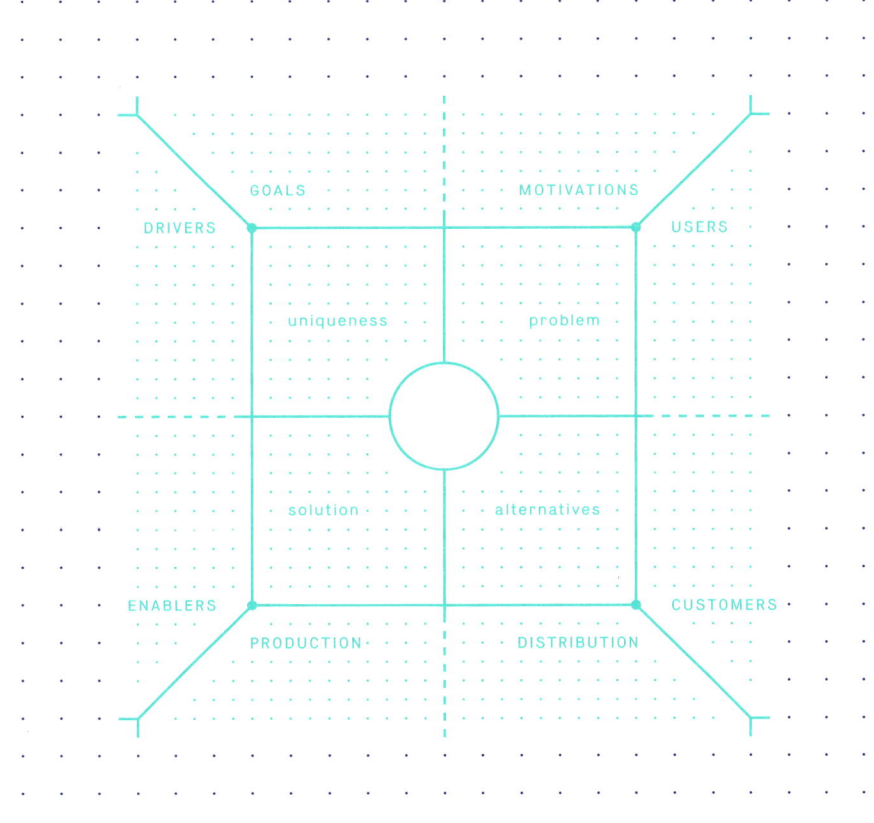

Fig. 15: Aspects of product innovation

Each visible feature is represented by a field describing its role and its place within the product's context. As for the realization axis, the more abstract and closer to the purpose of the product innovation, the further towards the top of the diagram it reaches. Conversely, the more concrete and relevant to implementation, the further down it can be found in the Product Field. The same applies to the introduction axis, where relevance to the company's internal organization puts the axis near the left *(inside)* and external aspects are located to the right *(outside)*.

2 MAP —
Big picture as a common working base

Mapping and collating the team's competencies in product innovation helps bring existing knowledge, as well as hidden expectations, to light. This gives a complete and structured picture that can be shared.

It would exceed the scope of this brief introduction to try and describe each of the 12 aspects of product innovation and their interdependencies in detail. For those who wish to dive deeper, we recommend the book *The Product Field Reference Guide* by Klaus-Peter Frahm, Michael Schieben, and Wolfgang Wopperer-Beholz, which can be viewed online or bought through the ProductField.com website. In this section, we will restrict ourselves to describing a few examples drawn from the bottom right-hand corner of the diagram and concerning implementation:

Customers

Customers are either individuals or organizations who expect to benefit from using the product and are therefore willing to pay. They can be, but do not necessarily need to be, users of the product themselves. To understand how they tick, it is important to know what motivates their buying decisions, where and how they can best be reached, and how much they may be willing to spend on the product. We suggest focusing on certain roles, industries, professional groups, company sizes, or addressability.

Guiding questions: Who are the people or organizations that actually pay for the product? Why are they willing to pay for the product? How do they reach their buying decisions?

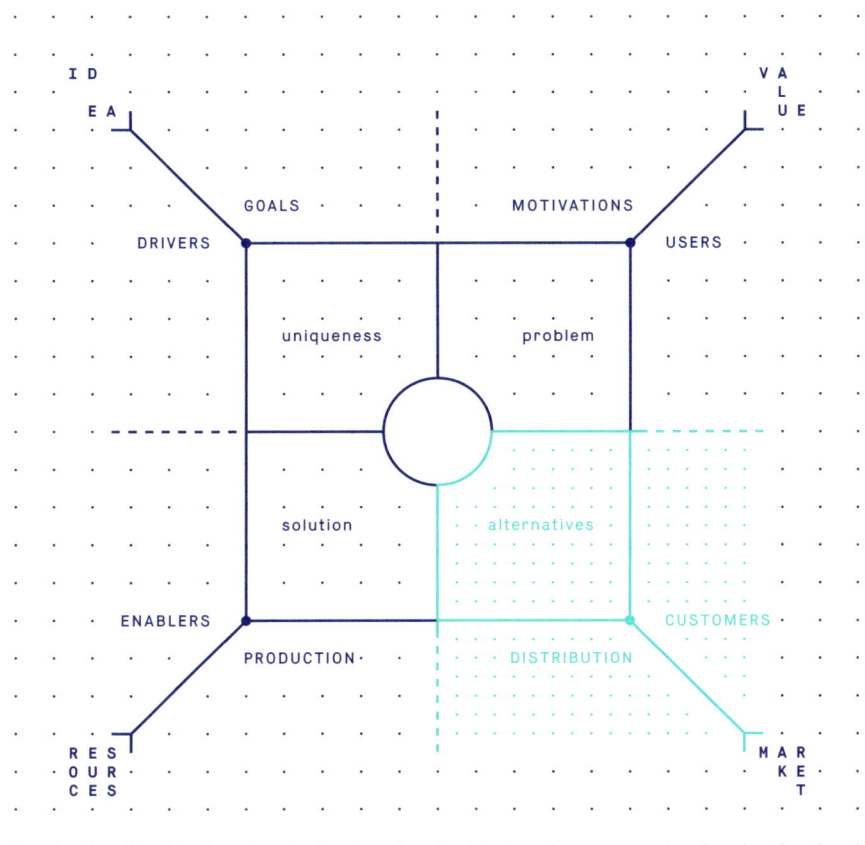

Fig. 16: Implementation / outside aspects

Distribution

Distribution is about bringing the product to the customer. This includes product marketing as well as customer acquisition and logistics. Understanding the intricacies of distribution for a given product provides important information that can affect go-to-market strategy and assist in determining customer

acquisition costs. It is the key to growth and ROI. Consider things like market, sales, logistics, channels, or technologies.

(?)   Guiding questions: Which channels, technologies, and partners can assist in bringing the product into the hands – and into the heads – of users and customers? What will it take to turn users or prospects into customers?

Alternatives

Alternatives are other solutions or processes already in use that currently address, or help to avoid, the challenge the new product is designed to overcome. These can stretch from competing products and services to the potential customer's inherent disinclination to address the problem. Alternatives are the benchmarks against which the uniqueness of an innovation, how many unique features the product offers compared to the alternatives, must be measured (see above $\longrightarrow$ 10x value) and grappling with them gives a realistic outlook about the product's chances in the market. It's a good idea to make sure all alternatives are known and have been studied. Don't just look at other products, but take any possible workarounds, DIY solutions, and sheer ignorance into account, too.

(?)   Guiding questions: Are there other ways to solve or alleviate the problem? What are users doing today to clear up the problem? Is it possible to ignore or bypass the problem?

Consult the book *The Product Field Reference Guide* for further information about aspects not discussed here, such as goals, drivers, uniqueness, motivations, users, problems, enablers, production and solution.

3   CHECK —
Find conceptual flaws

Experience shows that Transformational Products, especially in the early stages, are often plagued by conceptual shortcomings, gaps or inconsistencies. The faster we recognize this the better we can reconcile it, thus reducing the risk potential significantly. The Product Field is a good way to systematically assess concept quality by precisely locating aspects within the Product Field and following its inherent grammar.

Core/Context Fit

A good test of concept quality is how well does the product's value proposition fit the conditions and environments under which it originated and will be used. The value proposition is defined by four internal aspects, namely problem, solution, alternatives, and uniqueness. They form the core of any product innovation.

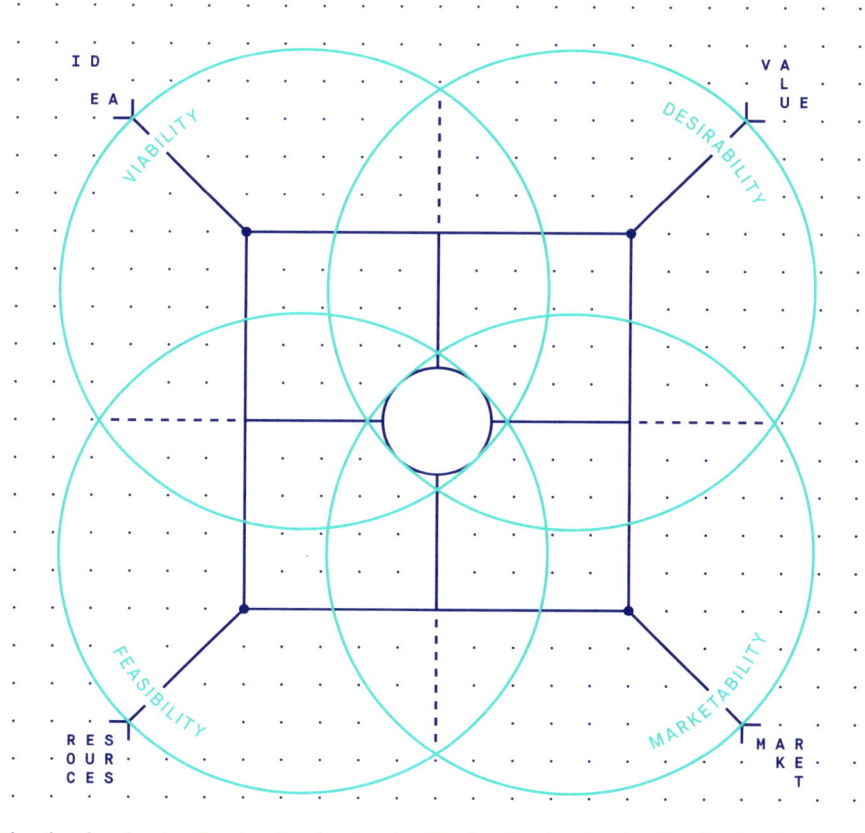

Fig. 17: The four qualities of core/context fits

This core is surrounded by aspects that have to do with the innovation's context. By their position within the Product Field it can easily be ascertained that each aspect of the core operates directly on its neighbors. This creates four "triangular relationships" that need to be both consistent and coherent. If so, we have achieved a perfect core/context fit: the value proposition conforms with the strategic idea behind the product, can be achieved within the framework of the given resources, addresses a sufficiently large market, and creates true value for the customer.

The four triangular relationships between a core and two context aspects are used to describe a product's → marketability, desirability, → viability, and → feasibility. If we are able to formulate a single sentence that describes the triangular relationship succinctly, then we can assume the concept is good enough.

Marketability

An innovative product will have a chance of succeeding if there is a big enough market for it and if the company has a distribution arm that is long and strong enough to reach a sufficient number of customers willing to substitute it for existing alternatives.

Use the following sentence template to check for marketability:

[Distribution] reaches [customers] willing to substitute [alternatives] with the product.

Desirability

An attractive product is one that really creates value for the customer or solves a problem that prevents customers from achieving some motivational result.

Use the following sentence template to check for desirability:

The product solves [problem] standing between [user] and his [motivation].

Viability

A product innovation will be sustainable if it is the expression of a practical business idea, if the product's uniqueness is compatible with the company's goals and drivers.

Use the following sentence template to check for viability:

The [goal] guides the [drivers] and enables the product's [uniqueness].

Feasibility

A product innovation is practicable if it can be realized with the resources at our disposal and if production capacity and capability is sufficient.

Use the following sentence template to check for feasibility:

The [enablers] empower [production] to build the [solution] offered by the product.

4   FIND —
Uncover hidden strengths and weaknesses, then execute!

Whether a product is practical or not depends to a high degree on the agents, strategists, artefacts and conditions involved in making it happen. In considering a product's strengths and weaknesses it is therefore necessary to view it in the proper context.

Any good strengths and weaknesses profile needs to be created incrementally: First, collect facts and assign positive or negative values to them, depending on their possible effects on the innovation and its future success. The sum of these values will be either positive or negative, indicating that the aspect in question is either a strength or a weakness.

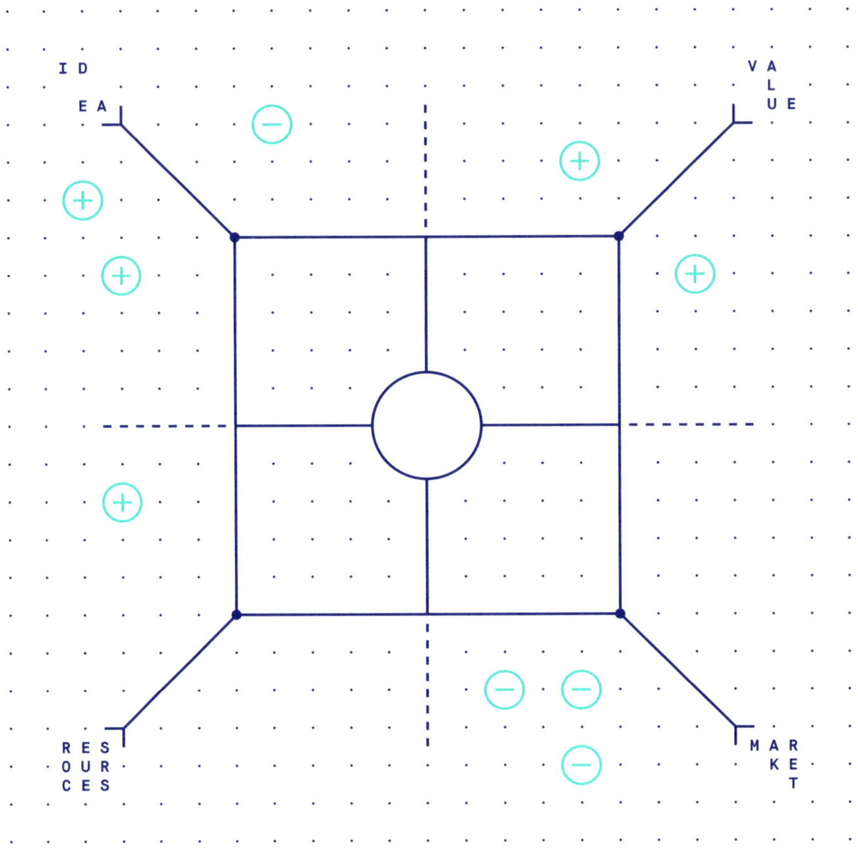

ID

EA

VALUE

RES
OUR
CES

MARKET

Fig. 18: Evaluation in context

Strengths

When looking for strengths it is always good to focus on benefits. For instance, clear goals, special assets, exceptional competencies, or easy access to certain market segments must be considered strengths.

Weaknesses

Knowing your product's weaknesses is an aid to risk management and optimization. Typical weaknesses include poorly-qualified actors or obstructionist team members, bad strategy, artifacts or conditions. Examples include weak design, conflicting goals, missing know-how, or blocked distribution channels.

Force field

Once the strengths and weaknesses of an innovative product have been found, the Product Field can be used to visualize combinatory forces. These are displayed along the introduction axis which represent the actual work which has been done on the project, for example, in its development and marketing. Forces can either propel a project outward, thus contributing to its success, or impede its introduction through disruptions or blockages. Promising product innovations can be easily recognized because their force fields move from inside to outside without obstruction, which means the product is delivering its promised value to its customers.

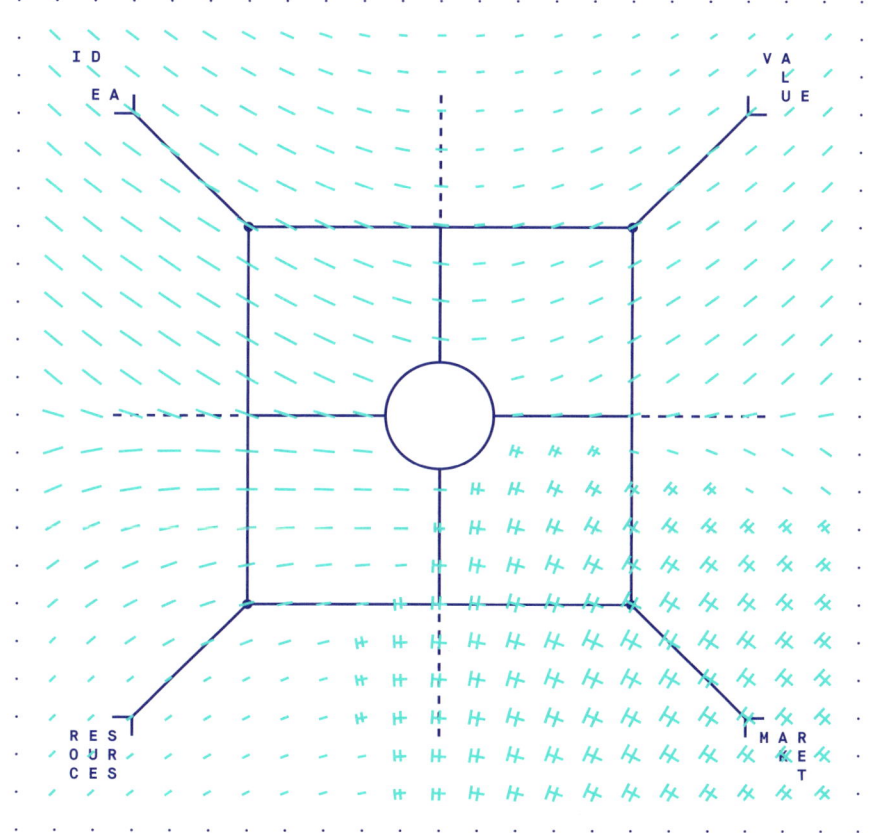

Fig. 19: Force field visualization

Strong opposition is a warning that a product is in present danger of failing. Management needs to focus on reducing or eliminating these opposing forces and thus the risk of failure. Existing strengths must be utilized to buttress positive forces and reduce the effects of negative ones.

HANDS–ON PRODUCT THINKING

There are two ways in which the Product Field adds practical value:

1. As a shared mental model, it helps the product team to focus and to communicate. Collaboration is improved, people tend to talk less over one another's heads, and they get to the point faster.

2. As a tool, the Product Field is a canvas that comes with a well defined four-step process to work with. Those steps are frame, map, check, and find. More information and practical tips can be found in *The Product Field Reference Guide* for those who need greater detail.

Product Field and product development methodology

The Product Field gives teams and organizations a good way to use existing tools and methodologies for product development more effectively and purposefully.

The gaps, weaknesses and risks, as well as the strengths and potentials discovered through the Product Field and by find lead to a number of very different challenges. These can be analyzed and contextualized together using the Product Field, which also enables the team to work out the right response in a variety of ways, for instance by using one or more of the following tools and methods:

- How might we
- Challenge mapping
- Jobs to be done
- Persona

- Empathy map
- Design studio
- Prototyping
- Context map
- Stakeholder map

Once a company realizes where it needs to go and what strengths and weaknesses its product innovation has – what its innovation profile looks like – it can then commence to either create a special innovation process or to beef up an existing one in order to find the right approach. The following strategies suggest themselves and can easily be mapped onto the Product Field:

- Lean start-up/customer development
- Design Thinking/Service Design
- Theory of constraints/Kanban

Product Toolbox

On the next double pages, you will find a series of methods and tools that have proven themselves in the past. Describing each in detail would go beyond the scope of this book. Instead, we will restrict ourselves to giving a suggested reading list for each.

DESIGN THINKING

Ever since the *Harvard Business Review* put it on its title page in 2015, Design Thinking has become the method of choice for most large enterprises seeking to innovate in the context of digital transformation. However, it is important to realize that Design Thinking is a process and a school of thought, not a single method. It sees innovation as the overlap between what is technically possible (feasibility), economically profitable (viability), and attractive to users (desirability).

Design Thinking talks about people, not users or consumers, showing just how broad the approach really is. It offers a universal method for solving problems and offers a range of process models to do so. Unlike the classical sequential "waterfall model" of designing, these processes are all non-linear and iterative. Each goes through a different set of phases, but progress doesn't have to follow a sequential course; the procedure may vary, and phases can be repeated.

This is because each phase can lead to insights that affect other phases. All they have in common is their user-centric focus. The big idea behind Design Thinking is that empathy for the user should not just feature theoretically in product development and be given lip service, but should instead be part and parcel of the process itself.

We will be introducing individual elements of Design Thinking such as (rapid) prototyping and testing later on in the toolbox section.

Suggested reading

→ **Brown, Tim** (2009).
Change by Design: How Design Thinking Transforms Organizations
and Inspires Innovation. HarperBusiness.
—
*One of the best entry-level books about Design Thinking containing lots of
examples, but lacking frameworks and worksheets. This is not a book written
by designers for designers, but caters to executives interested in making
Design Thinking an integral part of their enterprise and its products.*

→ **Kelley, Tom** (2001).
The Art of Innovation: Lessons in Creativity from Ideo, America's
Leading Design Firm. Crown Business.
—
The classic work from one of the inventors of Design Thinking.

→ **Kumar, Vijay** (2012).
101 Design Methods: A Structured Approach for Driving Innovation in
Your Organization. Wiley.
—
*A step-by-step introduction to Design Thinking. The author believes the
process that can lead to new products, services, and customer experiences is
not an art but hard science. He offers a range of practical tools and methods
for planning and defining new products.*

→ **Lockwood, Thomas** (2009).
Design Thinking: Integrating Innovation, Customer Experience, and
Brand Value. Allworth Press.
—
*A terrific overall view of the method, this book gives a detailed approach to
Design Thinking from a variety of perspectives as well as insights into the
challenges involved in bringing it into the company and making it part of a
business's processes and corporate culture.*

SERVICE DESIGN

Service Design and Design Thinking are so closely interrelated that some shrink them into one phrase as Service Design Thinking. In their book *This is Service Design Thinking*, which has become a classic, Marc Stickdorn and Jakob Schneider show 25 visual methods and tools they developed actually working with Service Design.

While Design Thinking may describe the use of certain design techniques to solve specific problems that may or may not be design-related, Service Design is a true design discipline, really a sub-discipline of product design. It defines services as complex, hybrid artefacts of things, locations, communication and interactive systems as well as people and organizations.

The service blueprint, a diagram that helps depict the complex nature of services and their transaction pathways (processes) is one of the oldest methods of visualization in Service Design. Meant originally to reflect the viewpoint of service providers, service blueprints have come a long way and now can be employed collaboratively or from a user-centric perspective. They can also be used to describe the underpinning services of the process. For everyday use, some ethnographic methods have also come to the fore, involving the close study and description of users, detailed analysis of the artefacts involved, and video diaries.

The overarching aim of Service Design is to provide insights into how people live their lives, how they acquire products, how they use them for their private and professional purposes. This calls for empathy, namely the ability to put oneself into another person's shoes and understand just how they would use the service. Co-designing is a crucial part of Service Design as it makes the user part of the collaborative development process, which can lead to important insights and successful results. Co-designing draws attention to often-neglected bits of knowledge, many of them by nature non-verbal, non-linear and intuitive.

Suggested reading

→ **Curedale, Robert** (2013).
Service Design: 250 Essential Methods. Design Community College Inc.
—

The author lists the various competencies a designer needs to master in order to succeed in developing services and experiences, and he gives numerous examples of promising methodologies.

→ **Osterwalder, Alexander et al.** (2010).
Business Model Generation: A Handbook for Visionaries, Game Changers, and Challengers. Wiley.
—

Not really a book about Service Design at all, rather a description of tried and proven methods that help in putting the principles of Service Design and Design Thinking into practice. The heart of the book deals with the business model canvas, a tool for defining business models from a strictly user-centric viewpoint.

→ **Polaine, Andy** (2013).
Service Design: From Insight to Implementation. Rosenfeld Media.
—

A great mix of theory and practice, this book offers a multitude of case studies that demonstrate how to put theoretical knowledge to work. It takes the reader through the whys, whats and hows of Service Design and clearly demonstrates their importance for users and their relationship to a service.

→ **Stickdorn, Marc et al.** (2010).
This is Service Design Thinking. Wiley.
—

A book for beginners that describes Service Design's interdisciplinary approach in plain, easy to understand language. It encourages readers to think about Service Design by describing five basic principles, demonstrating the similarities and differences between the various disciplines that make up Service Design. It outlines the iterative designing process and provides its readers with 25 flexible and adaptable Service Design tools.

PROTOTYPING

Traditionally, prototyping is a tool for exploring the user experience. It covers a wide range of methods that have one thing in common, namely their ability to transform theory into practical results. Prototyping creates visible products that can be experienced directly. It makes the product tangible for the very first time. It can be tested, and based on the results of these tests, an improvement process can be established. This is where a cycle sets in that comprises three steps or stages: prototype, review, and refine. This cycle is repeated until the prototype is finished and we can move on to construction. A prototype is not a → minimum viable product (MVP) in the true sense. A MVP can be used by real customers, while a prototype is only there to allow developers to experience a product's look and feel.

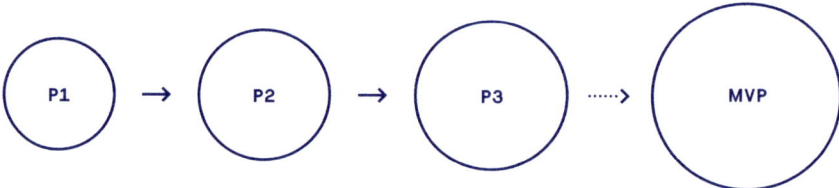

Fig. 20: From prototype to MVP

Typically, a prototype will start small and grow over its development in both breadth and depth until, ultimately, every aspect of the final product has been added. The Pareto principle (also known as the 80/20 rule) is a good one to follow here, focusing development on the 20 percent of the functionality that provide for 80 percent of usage time. Prototyping covers everything from simple sketches, that can be useful in the initial stages, to complex interactive simulations that, for all intents and purposes, look and work just like the finished product. We strongly suggest working with real code at the earliest possible moment. When starting with the MVP, we take the findings of the prototype and start with coding on a clean code base. That way we can improve the prototype fast and avoid to take over messy code. Prototyping stands and falls through the way users are made part of their development. They can provide critical feedback to trigger a learning process that is key at this point in product staging.

Suggested reading

⊕ **Gengnagel, Christoph et al.** (2015).
Rethink! Prototyping: Transdisciplinary Concepts of Prototyping.
Springer International Publishing.

—

A mainly academic textbook that claims products are multifunctional, inter-active systems, and therefore cooperation between the various disciplines is the only way to manage the resulting complexity. The authors argue convincingly that there are three basic ways to put the concept of prototyping to work in interdisciplinary teams.

⊕ **McElroy, Kathryn** (2017).
Prototyping for Designers. Developing the Best Digital and Physical Products. O'Reilly Media.

—

This book introduces readers to four methods of creating prototypes – from quick & dirty to hi-fi – and gives useful tips on how to test them with users.

⊕ **Nudelman, Greg** (2014).
The $1 Prototype: Lean Mobile UX Design and Rapid Innovation for Material Design, iOS8, and RWD. DesignCaffeine Press.

—

Prototyping especially for mobile; lo-fi prototyping with paper. The focus is on mobile, lean, and user experience (UX) design.

⊕ **Warfel, Todd Zaki** (2009).
Prototyping: A Practitioner's Guide. Rosenfeld Media.

—

A good book for beginners. The second chapter describes various tools in great detail, some of them unfortunately already obsolete.

DESIGN SPRINT

The first step is always the hardest, and nowhere is that more true than in product development. Kicking the process off and getting it up to speed takes lots of time and effort, both of which there never is enough. Jake Knapp, who works today as a design partner at Google Ventures (GV), has pressed the main elements of Design Thinking into a five-day regimen he calls the Design Sprint, a method which lets you move from problem description to a finished and user-tested prototype in a single week. But it isn't really about speed; it's about achieving momentum, focus and trust in the solution found. A Design Sprint helps the team orient itself and clearly define the goals it wishes to attain.

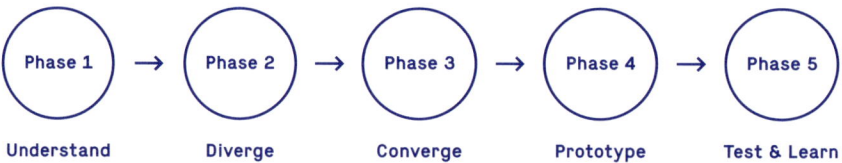

Fig. 21: Design Sprint

Typically, a Design Sprint consists of five stages, each to be run through in a single working day (see fig. 21). The objective is to take a product idea, move it to the prototype stage and perform tests designed to fill the team's knowledge gaps, validate or refute the riskiest assumptions and thereby put your project on the surest pathway to success. Let's take a closer look at the five stages:

① **Understand.** Develop a consensus of the problem, the business case, the customer/user, the value proposition, the success criteria, and the greatest risks. For this phase, we suggest working with a business model canvas.

② **Diverge.** Search for insights and possible solutions to your customers' problem. There are usually lots of ways to move from problem to solution and to define the goals to be reached through prototyping.

③ **Converge.** From the alternatives developed in the first two stages, choose the ones you want to pursue and that are within your scope. Develop a storyboard for the prototype and list all assumptions made along the way. Then draft a plan for testing these assumptions in hope of confirming them.

④ **Prototype.** Use the storyboard to construct a prototype that can be tested under real-life conditions and with real customers/users. Depending on circumstances, these can be realized in a number of ways including paper models, digital presentations, or simple HTML/CSS visuals.

⑤ **Test & learn.** Test the prototype with real customers/users and watch closely how they interact with it. Collect the results and use them to create a plan for further development.

A Design Sprint is a helpful tool to get the development process started and put the team on the right track. It creates a strong sense of commitment by essentially locking the entire team away together for a week in order to focus on the product and to conduct interviews with customers and users. It thrusts them forward from the realm of the abstract into real life and forces them to reach clear decisions incredibly quickly. Full of aplomb, the team can rise to new heights and move forward and achieve results faster than ever. A Design Sprint doesn't have to remain a one-off, it can be repeated over and over whenever the team reaches an important crossroad or when new features or elements need an injection of creativity.

Suggested reading

→ **Knapp, Jake et al.** (2016).
Sprint: How to to Solve Big Problems and Test New Ideas in Just Five Days. Simon & Schuster.

—

Essential reading for anyone interested in exploring the Design Sprint concept in more detail.

AGILE DEVELOPMENT

Within the developer community agile methods like → Scrum are already well-known. Like Design Thinking, Agile Development is the reaction to the failed linear methods of the past. Instead of artificially separating development stages like research, specification, and implementation and forcing them to stand in line, Agile Development teams let designers work together directly with product managers and designers in hopes of finding the perfect solution to the user's problem. Agile Development doesn't stand for a single method. Instead, agile teams pick the way they want to work and modify their methods when needed. Besides Scrum, the best-known methodologies today are probably Kanban and Extreme Programming (XP). Kanban's roots reach back to Toyota in the 1950s. It is a method for visualizing the flow of work, in order to balance demand with available capacity, and to spot bottlenecks.

All agile methods share short development and release cycles. This means that tests can be conducted faster and more often, which leads to more, and better, customer feedback. As a result, mistakes and wrong turns come to light quicker than with old-fashioned long-cycle development. Cycles – in Scrum development we refer to them as "sprints" (see Design Sprint) – typically take one to four weeks. In Scrum, the length of the cycles usually remains constant, while Kanban completely dispenses with fixed cycles. In the agile world, features are usually ready to use immediately once they are developed, without pre-defined release dates.

Suggested reading

⊕ **Skarin, Mattias** (2015).
Real-World Kanban: Do Less, Accomplish More with Lean Thinking.
Pragmatic Bookshelf.

—

Four case studies showing Kanban and Lean Thinking in action at three well-established companies where many obsolete systems, processes, organizational forms and old-school ways of thought are still to be found. Each study was under great pressure from the competition and needed to find ways of working more intelligently and how to produce results faster. Instead of turning to cost-cutting or reorganization, each achieved radical change in the way their teams work together, making it easier for them to deliver cutting-edge products.

⊕ **Stellman, Andrew et al.** (2014).
Learning Agile: Understanding Scrum, XP, Lean, and Kanban.
O'Reilly Media.

—

This book explains how agile methods work – why they are conceived the way they are, how they are deployed, which problems they solve, and which values, principles, and ideas they represent. Besides the approaches mentioned above (Scrum, XP, and Kanban) the book also discusses lean development.

⊕ **Sutherland, Jeff** (2014).
Scrum: The Art of Doing Twice the Work in Half the Time.
Crown Business.

—

The author is one of the inventors of the Scrum method. He describes the history and present state of agile methods in an extremely knowledgeable and often anecdotal way. He believes that Scrum methods can be used for much more than just software development.

LEAN

The idea behind the many branches of the lean method family is to eliminate everything that does not produce value and to focus instead on things that need to be done at a certain point in time. The lean approach works for companies of every size, from lowly startups to huge conglomerates. Like Kanban, the method traces its roots back to the legendary Toyota Production System developed in the late 1940s. Mary and Tom Poppendieck adapted the lean approach to software development. Steve Blank and Eric Ries ported it to the world of startups. Also known under the label Lean Innovation, these methods serve primarily as innovation tools.

We recommend lean methods for a number of reasons. For one, they place the focus firmly on the value created for the user. They also appear to be far more than just a passing fad. In many companies, lean methods have been learned over decades which makes them easy to apply to product teams and to adapt to existing product development processes. The methods work well with other approaches like Agile Development or Design Thinking, predestining them to use by interdisciplinary product teams.

Suggested reading

 **Blank, Steve** (2013).
The Four Steps to the Epiphany. K&S Ranch Press.
—
Originally published in 2003, this book puts the traditional process of product development on its feet. Steve Blank calls his concept "Customer Development", because he is quite right to focus on the user.

⊕ **Blank, Steve et al.** (2012).
The Startup Owner's Manual: The Step-By-Step Guide for Building a
Great Company. K&S Ranch Press.
—

*This can be read as a primer for aspiring product managers on the lookout for
a step-by-step guide to building a great product. The book combines customer
development with the lean startup approach.*

⊕ **Poppendieck, Mary et al.** (2003).
Lean Software Development: An Agile Toolkit.
Addison-Wesley Professional.
—

*This book explains traditional lean principles modified for software devel-
opment and introduces a set of 22 tools. A must-read for product managers,
managers in a technical leadership position, and anyone seriously involved
in product development.*

⊕ **Ries, Eric** (2011).
The Lean Startup: How Today's Entrepreneurs Use Continuous Innovation to Create Radically Successful Businesses. Crown Business.
—

*Eric Ries sticks to a strict logic that is a slap in the face for any would-be tech
mogul. Reis insists they should first test their ideas before betting the farm
on their latest brainchild. Don't listen to focus groups, he says, but instead
keep your eye on your customers and find out what they are really doing.
Start with a simple product and expand on what works. Expect failure and
be prepared to suffer through a painful and prolonged testing period until
you are absolutely certain you're on the right track.*

⊕ **Sehested, Claus et al.** (2010).
Lean Innovation: A Fast Path from Knowledge to Value.
Springer.
—

*The authors offer a set of principles that managers can use to supervise their
innovation processes better, and they discuss methods calculated to bring
result-oriented strategies and continuous learning to bear on innovation.*

TESTING

Possibly the biggest misconception around testing is the assumption that everything revolves around usability tests and functional code tests – so it only concerns hardware and software. That is undoubtedly important but it makes no sense to create a usable product if nobody wants to use it. Far more important, therefore, are questions like: Is there a market for the product? Do users love it? And will they eventually use it?

Don't wait until a product is finished before asking these questions and finding negative answers. By this time, making changes will really cost you and it is crushingly frustrating to have to go back and modify before you can launch. Making corrections early saves on time, money and nerves. All you have to do is start testing as early as possible, ideally during the lab stage. That way you can quickly find out if users are willing to make the product part of their everyday lives, whether it satisfies a real need, and what needs to be changed to make that happen.

Early user tests offer a way to add fast fixes during product development before costs start piling up and too much money has been invested in a flawed design. The risk of landing a flop later is also greatly reduced. Take time during the research phase to explore the problem range. This is where you need to make sure that the thing you are trying to solve really is an issue for your users, or whether your product is simply a solution in search of a problem.

Early testing will also show you how people have been dealing with the problem in the past. Do they use other tools? Have they developed some kind of workaround that creates the same results, namely to fix the problem or avoid it altogether? And how annoying do they think the problem is, anyway? As soon as you know for certain what the problem is that needs fixing, you can start to develop hypothetical solutions. By this time you should already be experimenting with various approaches to alleviating the problem, even if you don't have the perfect answer yet. Do user tests on the first prototypes with actual customers and generate user feedback at the earliest possible stage of development to avoid wasting time.

At this point, feedback from the market, not from friends and colleagues, will be needed – only truthful answers will serve now. Ask potential users whether they think the product fits their lifestyles. Keenly observe any emotion (or apathy) shown. As soon as you have a clear picture of the direction your product should be heading, use customer feedback to fine-tune the product concept, eliminate unnecessary features, and add anything that's deemed missing. Ask users what they would personally like to be different.

You can test prototypes very easily with tools such as InVision which lets users explore the product at their leisure, and by monitoring the behavior, you can quickly find out what needs to be done. Ongoing user tests are a good way to move beyond mere usability testing and create a product that people will want, love and, above all, use all the time.

Suggested reading

➔ **Hansen, Jared** (2015).
How to Jumpstart User Testing: 16 Tools to Craft Better Products. Silver Fox Marketing Group.
—
This handbook gives a short, precise overview of 16 different methods for conducting user tests.

➔ **Wolpers, Stefan** (2015).
Lean User Testing: A Pragmatic Step-by-Step Guide to User Tests. Berlin Product People.
—
Short, tightly-focused primer for anyone engaged in product development, from product managers to developers and designers (UX/UI). The book contains everything you need to know in order to start your first user tests as soon as possible.

LEAN ANALYTICS

Done the right way, analytics is one of the sharpest tools in the product developer's toolkit. As a rule, the analytics code should be ready by the time product code meets its first real user, at the latest by the construction stage. Analyzing user habits at the earliest possible stage reaps big rewards during later phases of development. Like testing, analytics helps to prove hypotheses, but while testing primarily produces qualitative feedback, analytics gives quantitative results, which is why the two go hand in glove.

Following the Lean Analytics approach means that the focus should be on metrics that trigger practical measures. A growing number of users might seem like a useful metric, but adding people won't make anything happen, whereas a sinking conversion rate is a real alarm signal that requires immediate attention. If users stop converting, you have to find out why, and then fix the problem because obviously something has gone wrong somewhere in the course of development.

If you chose to use Lean Analytics for product staging, then make sure there is a reliable metric available at every step currently being worked on by the product team. This involves setting concrete goals – a value or result that represents a real improvement and must be reached before things progress. This will inevitably lead to a new hypothesis which demands further tests or changes in the product to prove its worth. By measuring results, it is demonstrably apparent if the desired improvement has been implemented or not. Depending on these results, the cycle of metrics, hypothesis, experiment and action can start all over again.

Suggested reading

→ **Croll, Alistair et al.** (2014).
Understanding the Value of Lean Analytics: Use Data to Build a Better Startup Faster. O'Reilly Media.

—

The book provides a practical, step-by-step path from first idea to finished product. As a primer for practitioners, it is not just an assist for startups because it provides more than 30 case studies based on interviews with more than 100 founders and investors.

→ **Foreman, John W.** (2013).
Data Smart: Using Data Science to Transform Information into Insight. Wiley.

—

The author explains the concepts behind data science methodologies by offering concrete examples. A good book for managers in need of a basic understanding of Lean Analysis.

→ **Kaushik, Avinash** (2009).
Web Analytics 2.0: The Art of Online Accountability and Science of Customer Centricity. Sybex.

—

A beginner's guide offering an overview of best practices for most areas in which analytics is involved.

Product Factory

Developing digital products is an essential part of the innovation business. This goes all the more for products aimed at being transformational. Individual innovation bets may fail for a company, but in the end the important thing is a balanced ledger. Among classic industries, pharmaceuticals, aka Big Pharma, probably understands this best. Here, failure is all part of the job. The important thing is to fail in the right places and in the right way. What can we learn from this?

Drug companies are paranoid by nature because they constantly have an appointment with death. Their blockbuster products, the ones that bring in the lion's share of revenue, are protected by patents that expire after 15 years. After that, the market will be swamped by cheap generic products. On average, Big Pharma has about eight years to make a killing because by then they will be cut off from their big money makers – unless, that is, they manage to create a strong product pipeline that keeps the moolah rolling in.

As we saw in the first section of this book, Moore's Law of hardware inflation in combination with network effects from the software side has spawned a growing influx of disruptive products. The drug business has been dealing with problems like those caused by digital markets for ages. Today's products aren't those of tomorrow, and they are dying faster all the time. Pharmaceutical companies teach us several important principles that apply directly to digital product innovation:

①   **Build a strong pipeline.** For drug companies, most bets on new products never work out. They invest tons of money, often billions, in new products and then watch their money run down the drain. Often, new medicines fail to pass the necessary stringent clinical tests. Failure is part of pharmaceutical's DNA, and the huge pressure on their product teams could easily lead them to fudge some of the results. However, the blow-off would probably be so horrible for both patients and a company's image that they usually stay in line. To stave off the uncertainty, most companies operate with a big portfolio of parallel products. Scientists working for Big Pharma shovel loads of product ideas into the funnel and transform them into a robust product pipeline by following strictly supervised development processes. That way, the probability is high that at least a handful of them will turn out to be the blockbusters for the markets of tomorrow.

2. **Invest big in the early stages of research and lab work.** Pharmaceuticals invest disproportionately in the creation of a big pipeline of products. Translated into the language of digital products, this means it's better to discover the right product than to make it the right way. It's no coincidence that the drug and biotech industries spend more of their earnings on R&D than any other. In 2014, Big Pharma spent 14.4 percent of turnovers on developing new products. That is the only way they can prime their pipelines with sufficient potential products to be sure that enough of them will turn up trumps. By comparison, the automotive industry only spends about a third on R&D (4.4 percent in 2014).

3. **Innovation management triple-play.** Doing their own research is considered very important by all the big players, but they are open for other sources of innovation, too. Here, digitization has been a real game-changer. For instance, thanks to the revolution in CRISPR/Cas9 genome editing techniques, it is as simple to reorganize a human genome today as it is to cut and paste snippets of text in a word processing program. This means that armies of third-party developers – private and contract researchers, as well as successful startups – offer huge market potential.

4. **Managing insecurity.** Pharma companies are past masters when it comes to dealing with uncertainty. They stick to strict process guidelines and work tirelessly on filling the pipeline. They also operate with hard gates between development stages to avoid costly aberrations. That way, they can concentrate their resources on a tiny handful of really promising product innovations.

⑤ **Scale and market fast.** In bringing pharma products to market, time is precious. Pharmaceuticals are very good at pushing new products into the global market once they have passed their clinical trials and been approved by the health authorities. Product development, certification, and marketing work together like clockwork, and departments and teams for each phase operate within different cultural environments.

These five principles will provide the blueprint for the last stage of our playbook. Once the final greenhouse stage of the Transformational Product is complete, the product team passes it on to the "product factory." By then it will have proved itself in a pilot market and passed the test of scalability.

The product factory's job is now to industrialize the product for various markets and user groups. We believe that large organizations can only differentiate themselves through Transformational Products that demonstrate the company's future viability in the eyes of its own employees. The company, therefore, proves itself as a product factory by demonstrating its ability to industrialize successful Transformational Products. The model for this is the → two-speed organisation: on the one hand, fast, agile product teams with mindsets of their own, relying heavily on trial and error when creating new products; on the other, a robust product factory with a proven track record for turning out successful products with the help of reliable, scalable processes and incremental improvements. Both have their justification and their strengths within the company, and both need to work together to create a resilient organization.

In the end, a steady stream of Transformational Products will redefine product categories and transform markets and, by way of feedback, transform the entire company. Fig. 23 shows how this principle applies to Amazon, Uber and Tesla.

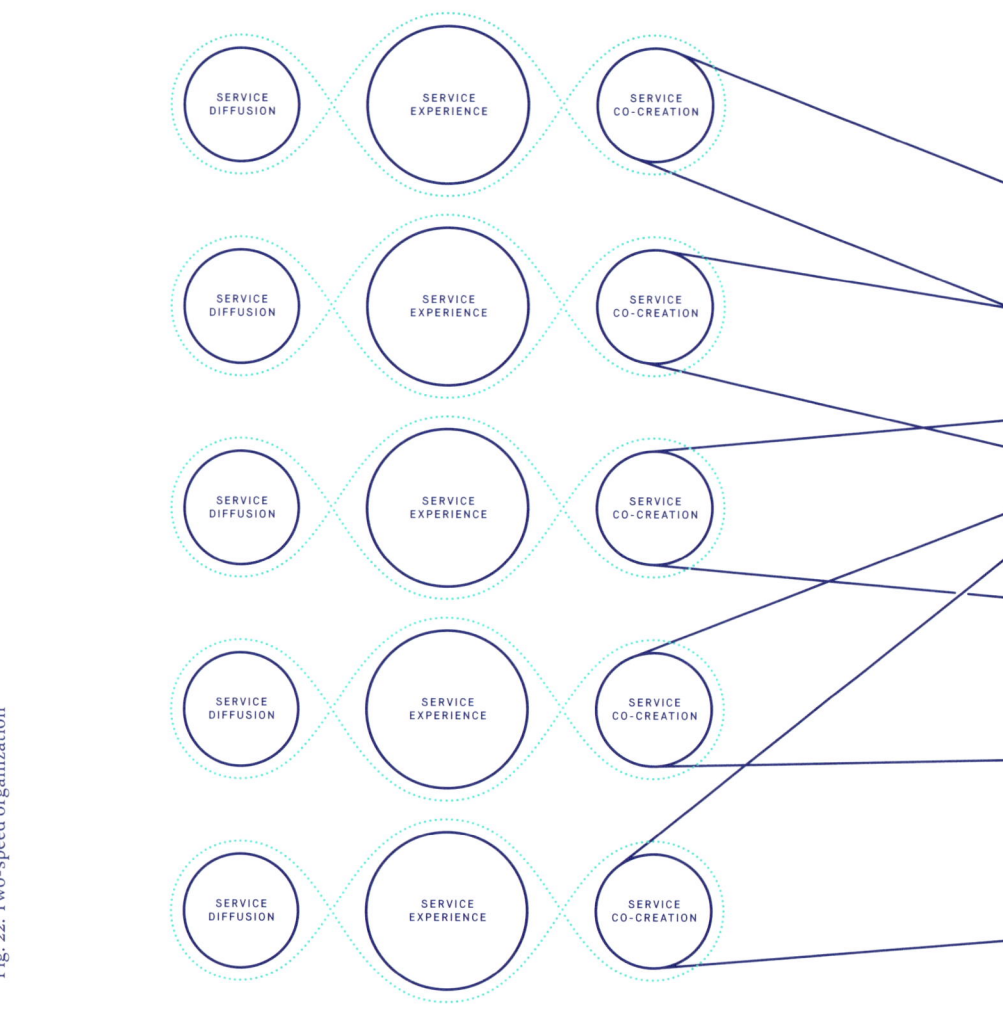

COMPANY TRANSFORMATION

Industrialization

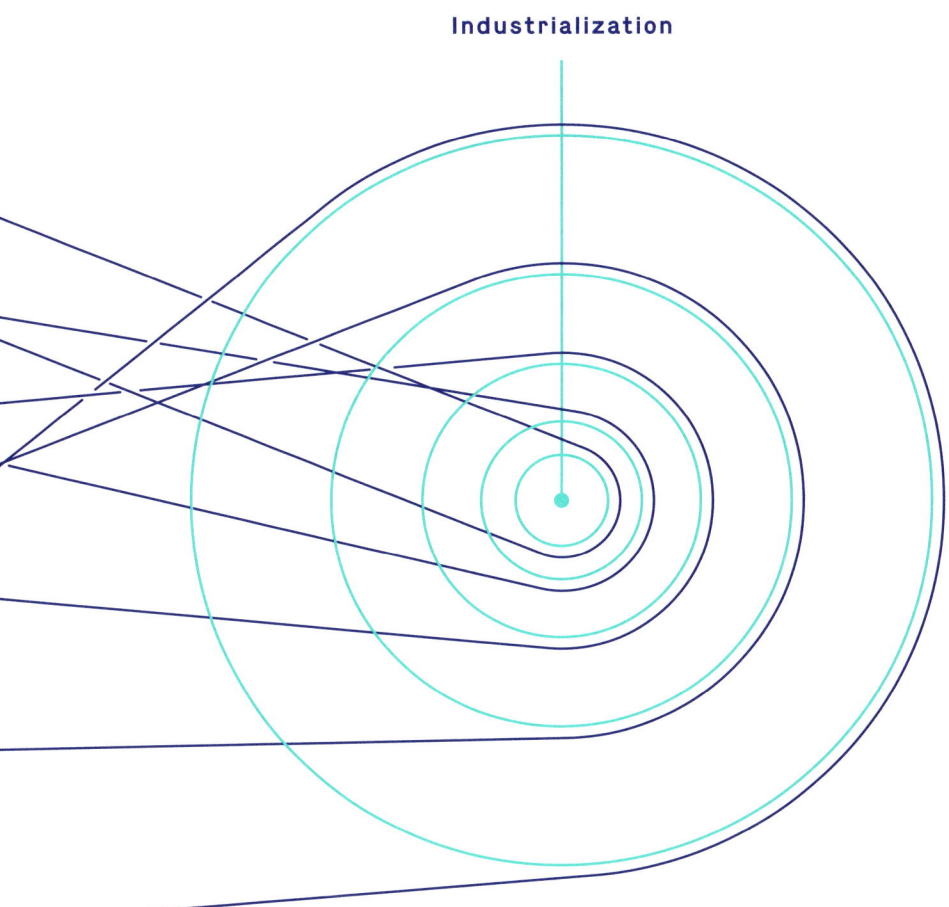

PRODUCT FACTORY: "Doing things the right way"

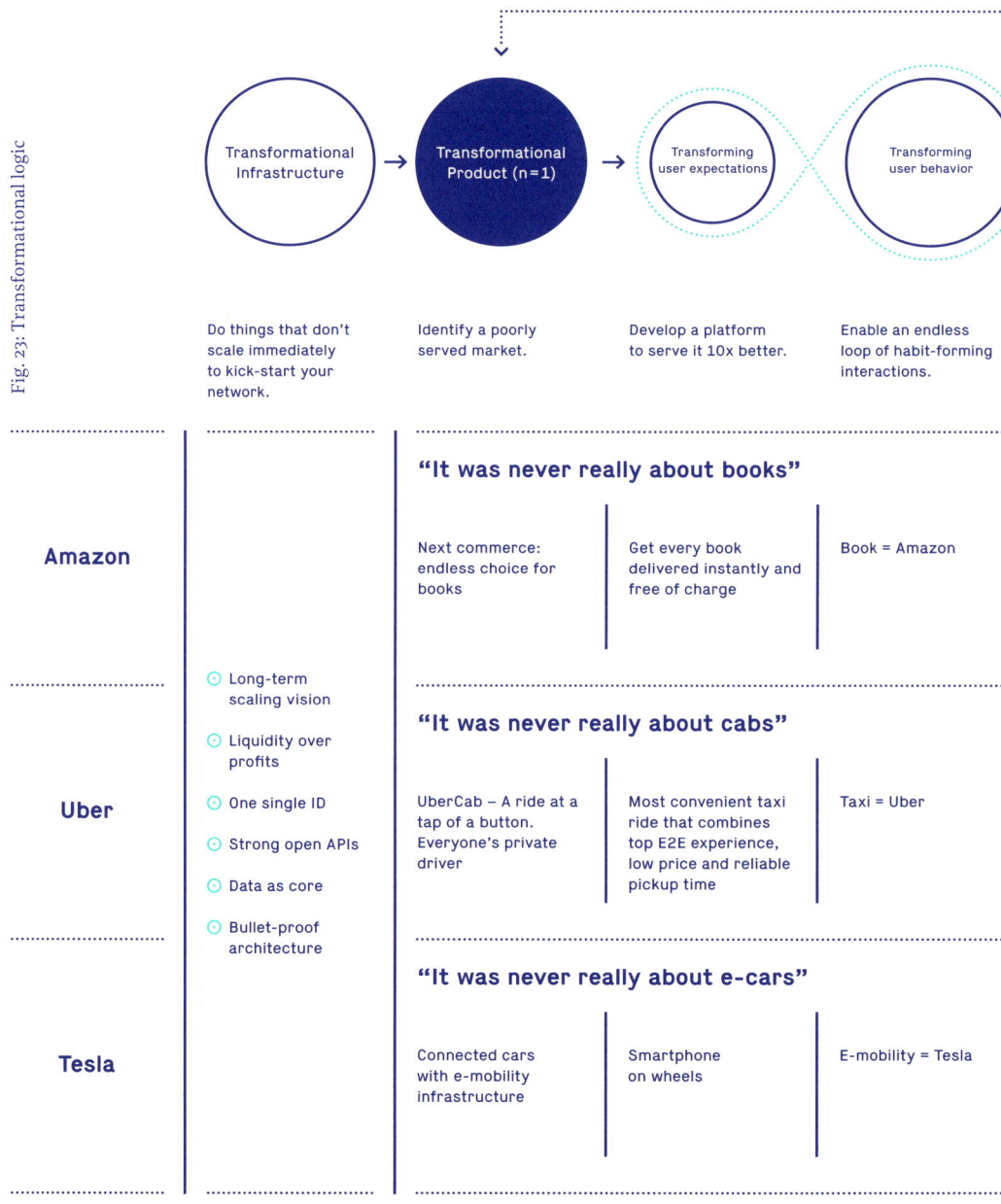

PRODUCT CREATING

Fig. 23: Transformational logic

Transformational Infrastructure → Transformational Product (n=1) → Transforming user expectations   Transforming user behavior

Do things that don't scale immediately to kick-start your network.

Identify a poorly served market.

Develop a platform to serve it 10x better.

Enable an endless loop of habit-forming interactions.

Amazon

"It was never really about books"

Next commerce: endless choice for books

Get every book delivered instantly and free of charge

Book = Amazon

○ Long-term scaling vision

○ Liquidity over profits

○ One single ID

Uber

"It was never really about cabs"

UberCab – A ride at a tap of a button. Everyone's private driver

Most convenient taxi ride that combines top E2E experience, low price and reliable pickup time

Taxi = Uber

○ Strong open APIs

○ Data as core

○ Bullet-proof architecture

Tesla

"It was never really about e-cars"

Connected cars with e-mobility infrastructure

Smartphone on wheels

E-mobility = Tesla

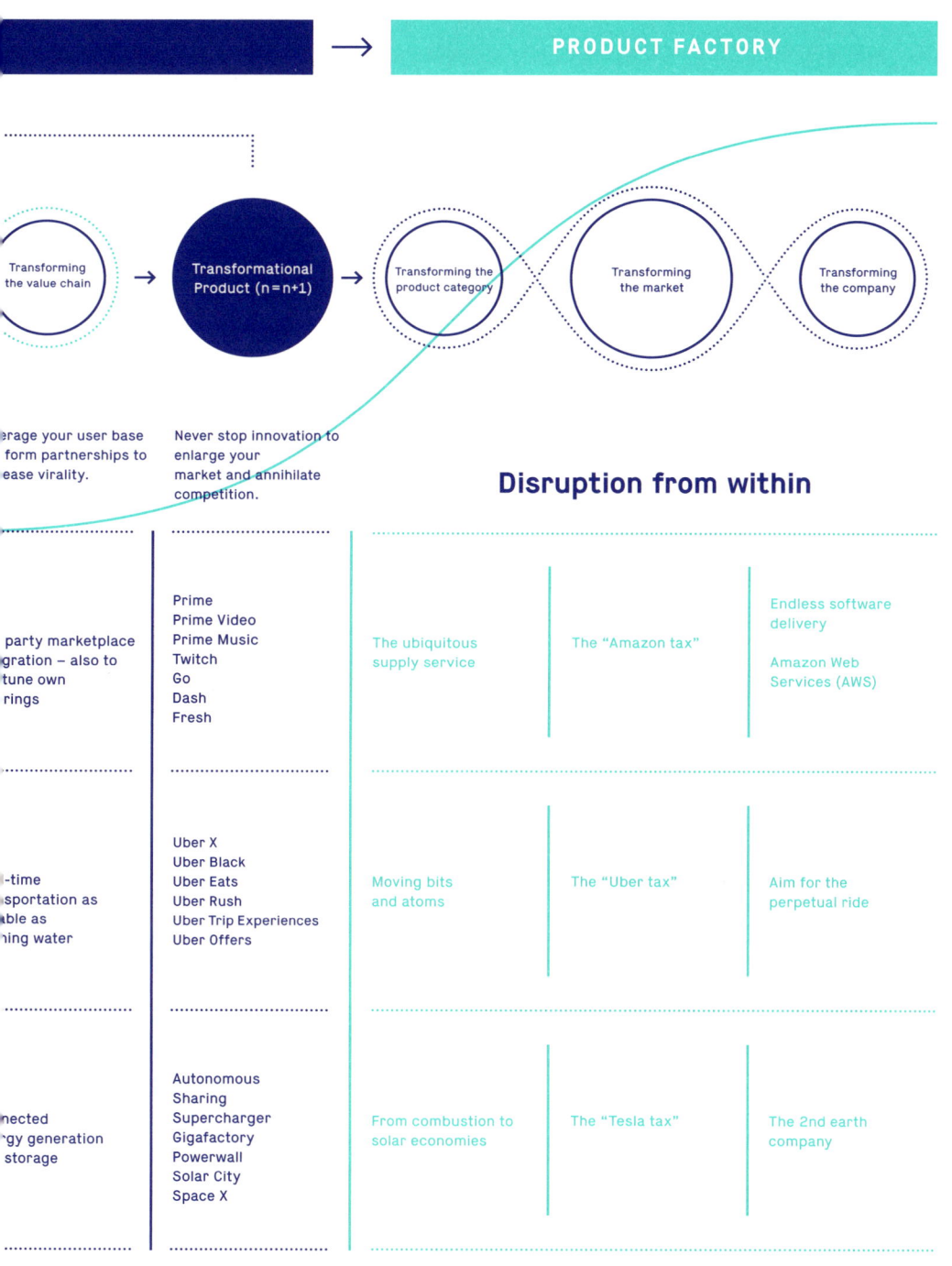

Transforming the value chain → **Transformational Product (n=n+1)** → **Transforming the product category** → **Transforming the market** → **Transforming the company**

...erage your user base ... form partnerships to ...ease virality.

Never stop innovation to enlarge your market and annihilate competition.

Disruption from within

...party marketplace ...gration – also to ...tune own ...rings	Prime Prime Video Prime Music Twitch Go Dash Fresh	The ubiquitous supply service	The "Amazon tax"	Endless software delivery Amazon Web Services (AWS)
...-time ...sportation as ...able as ...ing water	Uber X Uber Black Uber Eats Uber Rush Uber Trip Experiences Uber Offers	Moving bits and atoms	The "Uber tax"	Aim for the perpetual ride
...nected ...gy generation ...storage	Autonomous Sharing Supercharger Gigafactory Powerwall Solar City Space X	From combustion to solar economies	The "Tesla tax"	The 2nd earth company

EPILOGUE

Epilogue

We have already examined the powerful undercurrents shaping the Casual Economy: how pervasive technology and the platforms based upon it, created by the likes of Google, Apple, Facebook, and Amazon, have changed our lives. We asked why it is so important for corporate survival to focus entirely on developing a customized digital product pipeline. We have also shown how business IT, within a single generation, has been transformed into personal computing and why connectivity is at the core of digitization.

In the 1990s, early adopters of the internet spent minutes a day online; now our smartphones are always connected, 24x7x365. The net has become part of our daily lives and, on average, users glance at their smartphones 150 times a day, using them to perform tasks perfectly tailored to fit their immediate needs, desires, and whims. We use smartphones to organize, and sometimes even to replace, our social lives and to pass our time. Perhaps most importantly, our personal consumption is increasingly controlled through these tiny touchscreens. Three swipes of a fingertip can be sufficient to order some bright new gadget, rent a film, or call a cab.

For companies, however, this is not a big walk in the park. To reach new or existing customers, they have no alternative but to go through the colossal crossroad platforms of the internet. The digital Pure Players, above all the giant GAFA companies, have become the gatekeepers between companies and customers. The GAFA club is busy skimming the cream off the value created by companies all over the world and taking their cut of almost every kind of transaction imaginable. No wonder their market capitalization makes them some of the most valuable corporate entities on the planet.

If not to be crushed in the bear hug of the big platform operators, companies need to develop products that make them less beholden to GAFA and their ilk – products that integrate tightly and without causing friction into the everyday lives of millions. Lots of things currently being labeled as "digital transformation" deal with culture and processes. Myriads of consultants bend their clients' ears with tales of Californian startups. Sneakers and jeans have replaced suits and ties as the dress code of business; flat hierarchies and agile methodology have become the new mantras of management.

Not that they're wrong, of course; much of it, however, is pointless. Yes, we need to update our processes and create new mindsets within our companies, but that will not be enough to be successful in the Digital Age. By all means, go off on a guided tour of Silicon Valley, that new Disney World for executives, and admire the laid-back work styles at Google, Facebook or Uber – just don't fool yourself into believing that therein lies the secret of their success.

In reality, → unicorns are very elusive beasts, and the number of startups that have managed to boost their stock value above a billion dollars are rare. Most hopeful, young companies flounder even though they strive to emulate the GAFA culture and employ the same methods and processes as these paragons. You will seek in vain among those failed startups for waterfall methods, closed silos, and iron-bound hierarchies. If they failed, it wasn't for lack of open-mindedness and willingness to accept change. In fact, if you look more closely at management styles followed by the GAFA firms, you will find they are remarkable more for their diversities than for their similarities. It turns out there is no "digital corporate culture", a best practice blueprint you just have to emulate to be a winner is simply an illusion fostered by PowerPoint and Keynote presentations.

The future doesn't lie in copying the methods and the corporate habitus of a few Silicon Valley startups. Instead, it's important to create the right products – digital products with a chance of becoming the next blockbusters and thereby generating the revenues and profits needed to survive and prosper. Culture alone won't pay your salary.

Cracking the innovation dilemma remains a challenge because corporations are bred to reach for change incrementally, step by cautious step, focusing all the while on gaining efficiency. A large company's immune system is also used to destroying anything that could endanger the status quo, anything that could jeopardize jobs and careers is anathema.

Add to that the uncertainty that any given product innovation will succeed in the market, and it's understandable that many prefer post-rational explanations (which is like driving forward by looking in your rear-view mirror). Was it really inevitable that Google would come out on top? There were lots of other search engines, after all. And who would have thought that a tiny bookseller from Seattle would rise to become Amazon, the biggest retailer in the world? Why did Facebook outdistance the many social networks that came before?

Determining why a certain product succeeds and others don't has lots to do with timing and (whether we like to admit it or not) with serendipity. There is simply no way to tell in the beginning if the product under development will one day be worth a fortune or not. Only a handful of the thousands upon thousands of startups that are born every year really change the world, and venture capitalists are akin to roulette players. Big corporations, on the other hand, don't play dice. Careful planning and risk management are more their game. That's why, in the context of this book, we chose not just to discuss the characteristics and the code of Transformational Products, but also to include our playbook in part three which shows how individual products as well as parallel product initiatives emerge and flourish.

By focusing primarily on the product, organizations are also forced to place emphasis on the value the product creates for their customers. If this is done correctly the product's chances of success grow and with it the opportunity to change market behaviors and, consequently, to transform the company itself. Every enterprise seeks to find out how their products and services resonate in the market. And, if it becomes clear where the future of the company lies, career paths are affected, too, with everyone benefiting by conforming to the new pattern.

At SinnerSchrader, we have been developing digital products for over 20 years. We started out working with tiny startups like Intershop, Ricardo,

Libri.de and buecher.de, many of whom soon rose to become household names, executing very successful IPOs. Today, our clients are mainly large Fortune 500 or DAX enterprises. One thing we've learned over time is that sometimes you need to act radically. Radical comes from radix, the Latin word for root, and in business the roots of change lie in creating products that add value. Only these can truly transform the habits of individuals, of markets, and of entire enterprises.

Developing Transformational Products is hard work. It's non-linear, unplannable, often frustrating and demands making lots of tricky decisions along the way. But we believe very firmly that this is the right way to do things because it offers the best chance of success in the digital world. Hopefully, this book will be of help.

GLOSSARY

Glossary

agile: describes iterative methods of developing software and products through a series of short release cycles such as → Scrum, Kanban or Extreme Programming (XP).

application programming interface (API): connection used in applications development that makes a product extendable by third-party developer.

artificial intelligence (AI): intelligence exhibited by machines.

Atomic Design (AD): a methodology invented in 2016 by → Brad Frost that involves breaking a website layout down into its basic components, which are then reused throughout the site. AD consists of five stages: atoms, molecules, organisms, templates, and pages.

average revenue per user (ARPU): total revenue divided by the number of users and/or subscribers.

botnet: an often large number of Internet-connected devices, each of which is running one or more web robots, or bots. Botnets can be used to perform → distributed denial-of-service attacks (DDoS), steal data, send spam, or allow the attacker access to the device and its connection.

canvas: a diagram used to develop new or to document existing business models. It provides a window with elements such as the value proposition, infrastructures, customers, or revenue streams of a company or a product.

casual: fortuitous, passing, informal, irregular.

Casual Economy: process by which business is transformed through the introduction of convenience services, especially those operated by platform creators such as Google, Apple, Facebook, and Amazon (→ GAFA).

co-creation: co-creation: the idea that the value of a product cannot be thought of without including the user as a co-creator. It is considered one of the pillars of → service-dominant logic (S-D).

co-creator: a participant in → co-creation

consumerization: describes the proliferation of computers, which used to be the exclusive domain of experts but are now increasingly used by the average consumer.

conversion rate (CR): percentage of users who take a desired action, for instance ordering or buying something. The CR is an important measure with which to gauge the success of an ad campaign or the efficiency of a website.

cost per acquisition (CPA): also known as pay per acquisition (PPA) or cost per conversion (CPC). CPA is part of the online advertising pricing model where the average cost to the advertiser of each transaction is evaluated. For example, a sale, a click, or a form submission such as a contact request, a user registration or a newsletter sign-up.

customer journey: the → user experience of a web product over various touchpoints through which the user interacts with the product or its manufacturer. It spans the navigation path from initial contact to the completed, or aborted, sale.

customer journey mapping: graphical depiction of the → customer journey.

deep learning: branch of → machine learning based on algorithms in multi-layer networks capable of teaching itself using large amounts of structured and unstructured data.

deep packet inspection: form of computer network packet filtering where the header describing the data packet contents and/or the data itself is examined as it passes an inspection point. In the context of online advertising, it is used to collect information, typically through an Internet Service Provider (ISP), which can be used to choose suitable targeted advertising displays based on discovered web activity. Often used by national governments to monitor and control internet usage by citizens (→ Great Firewall of China)

Design Thinking: universal method for problem solving through design. It puts empathy with the customer in the center of design.

desirability: the quality of being attractive and/or worth having.

digital transformation: the ongoing digitization and its effect on consumer habits and the markets related to them as well as on the way companies organize. In a wider sense the transformation of society as a whole in the Digital Age.

distributed denial-of-service (DDoS): a cyber-attack where the perpetrator seeks to make a machine or network resource unavailable: typically accomplished by flooding the target with superfluous requests in an attempt to overload it and prevent some, often all, legitimate requests from being fulfilled. Usually involves a multitude of compromised computers working in unison (→ botnet).

edge case: a problem or situation that only occurs at extreme (maximum or minimum) operating parameters.

end-to-end: in development, this describes all phases of the design process. A product team with end-to-end responsibility is accountable for the entire design.

feasibility: the degree of being → feasible

feasible: possible and practical to do easily or conveniently.

fidelity: a prototype's degree of closeness to the final product. Described as low fidelity (distant), abbreviated to lo-fi, and high fidelity (close), or hi-fi.

FMCG: Fast-moving consumer goods.

full-stack: Used to describe developers who can work with both backend and frontend systems and masters of database development, tools like PHP, HTML, CSS, JavaScript, and other related technologies. In an innovation context, full-stack describes a product team capable of dealing with every aspect of the design process, including software and hardware design, marketing, supply chain management, sales, partnerships, regulation, etc.

functions-on-demand: activation of additional features in existing hardware or software on request, often for an additional fee. For example, extra security features in antivirus software or automatic software updates.

GAFA: collective acronym for the Big Four players in the digital world, Google, Apple, Facebook, and Amazon.

gate: a milestone in the product development process (→ stage-gate).

Great Firewall of China: term used sarcastically for the combination of legislative and technological actions that have been taken by the government of Mainland China to constrain website access domestically.

growth hacker: → growth hacking

growth hacking: a process of rapid experimentation across marketing channels and product development to identify the most effective and efficient ways to grow a business. In coding, it describes integrating marketing mechanisms into the product itself.

hacking: the act of breaching defenses in a computer system and networks. In a wider sense it describes the ability to penetrate well-defended markets.

Internet of Things (IoT): the connection of physical objects and machines with the internet, aiming to create a totally networked world.

living style guide: a living piece of code, acting as a reference for a team of designers and developers to understand how an application will look and feel. Also used to convey a design language.

lock-in: snap point where a digital product becomes an essential part of the customers life. We distinguish between mental lock-in where the user starts unconsciously performing certain tasks, and functional lock-in that tie users to a certain product.

machine learning: a subfield of computer science that gives computers the ability to learn without being explicitly programmed by drawing on large amounts of data (→ artificial intelligence).

marketability: a product that is readily saleable to a large target audience.

minimum viable product (MVP): As opposed to a prototype design only to visualize a product, an MVP offers basic essential functionality but can already be used by real customers.

monthly active users (MAUs): the number of unique users per month.

Moore's Law: an observation made by Intel co-founder Gordon Moore in 1965 stating that the number of transistors per square inch on integrated circuits doubles every two years.

pivot: point of radical change of direction during product development or within a company.

platform: a virtual (often cloud-based) marketplace where consumers and producers can meet and exchange services. As a business model, platforms employ technology to connect people, organizations, and resources through an interactive ecosystem.

priming: in psychology, priming describes a way of activating particular representations or associations in memory just before carrying out an action or task. Think Pavlov's dog.

Product Thinking: the ability to focus exclusively on the product under development (as opposed to thinking in terms of projects or processes).

real-time bidding: buying and selling of online ad impressions through real-time auctions that occur in the time it takes a webpage to load; the winning bidder's ad is then loaded nearly instantly.

Scrum: a form of → agile software development that involves short, constant development and release cycles. Originally used in software development, but now applies to other development processes as well.

service-dominant logic: a meta-theoretical framework for explaining value creation, through exchange, among configurations of actors.

stage/phase gate: a process in project management where new product creation, software development, process improvement, or business change is divided into distinct steps separated by decision points.

stakeholder alignment: a buzzword often heard in change management and project management circles, it describes unity of purpose among key decision makers and other relevant players.

sunk cost fallacy: a key concept in behavioral economics referring to costs that have already been incurred, but which cannot be recouped. The fallacy lies in honoring sunk costs instead of simply ignoring them, which would make more sense decision-theoretically.

tipping point: the critical point in a situation, process, or system beyond which a significant and often unstoppable effect or change takes place.

tweaks: a way of continuously improving a design by making fine adjustments to it.

two-speed organization: an organizational concept whereby → agile, innovative initiatives should be allowed to move forward quickly without being hampered by the checks and balances that are needed to maintain critical business processes.

unfair advantage: a way of achieving superiority that cannot be copied or purchased.

unicorn: a startup company valued at over $1 billion.

usability: the art of making products and systems easier to use and matching them more closely to user needs, requirements and abilities.

use case: demonstration of a typical application for a specific product, service, or technology.

user-centric design (UCD): a framework of processes in which usability goals, user characteristics, environment, tasks and workflow of a product, service or process are given extensive attention at each stage of the design process.

user experience (UX): emotional result of using a product.

user experience design (UXD): the process of enhancing user satisfaction with a product by improving the usability, accessibility, and pleasure provided in the interaction with the product.

user interface (UI): the space where interactions between humans and machines occur.

utility: the total satisfaction received from consuming a good or service.

value in exchange: the ability to trade an asset, such as money, for goods and services (as opposed to → value in use).

value in use: the value that a product generates for a specific owner under a specific use.

vaporware: a product which is announced and/or is being developed but never gets released or cancelled.

viability: the capacity of a product to operate or be sustained economically.

walled garden: an environment that controls the user's access to web content and services approved by the platform operator. Walled gardens are the opposite of open platforms where users generally enjoy unrestricted access to applications and content.

SOURCES

Sources

Ahmed, Ajaz et al. (2012). Velocity: The Seven New Laws for a World Gone Digital. Vermilion. NEXT

Andreessen, Marc (2007). The three kinds of platforms you meet on the Internet. blog.pmarca.com/pmarchive.com.

Andreessen, Marc (2011). Why Software Is Eating The World. Wall Street Journal.

Averdung, Axel (2013). Erfolgreiches Management von Marketingagenturen im Wandel: Differenzierende Kompetenzen als strategischer Wettbewerbsvorteil. Springer Gabler. NEXT

Bard, Alexander (2012). The Futurica Trilogy. Stockholm Text. NEXT

Bard, Alexander (2012).The Internet Revolution. nextco.nf/2m6PmdM. NEXT

Blaase, Nikkel (2015). Why Product Thinking is the next big thing in UX Design. medium.com.

Blank, Steve (2013). The Four Steps to the Epiphany. K&S Ranch Press.

Blank, Steve (2013). Why the Lean Startup Changes Everything. Harvard Business Review, May 2013.

Blank, Steve et al. (2012). The Startup Owner's Manual: The Step-By-Step Guide for Building a Great Company. K&S Ranch Press.

Brown, Tim (2009). Change by Design: How Design Thinking Transforms Organizations and Inspires Innovation. HarperBusiness.

Butlitsky, Michael (2013). The World is a Product. MindtheProduct.com.

Cagan, Marty (2008). Inspired: How To Create Products Customers Love. SVPG Press.

Cardone, Grant (2011). The 10x Rule: The Only Difference Between Success and Failure. Wiley.

Choudary, Sangeet Paul (2015). Platform Scale. Platform Thinking Labs.

Choudary, Sangeet Paul et al. (2015). APIs and Platforms: How Interfaces and Access Enable the Networked Economy. platformed.info.

Christensen, Clayton (1997). The Innovator's Dilemma. HarperBusiness.

Christensen, Clayton (2003). The Innovator's Solution: Creating and Sustaining Successful Growth. Harvard Business School Press.

Cooper, Robert G. (2001). Winning at New Products. Basic Books.

Croll, Alistair et al. (2013). Lean Analytics: Use Data to Build a Better Startup Faster. O'Reilly Media.

Curedale, Robert (2013). Service Design: 250 Essential Methods. Design Community College Inc.

DIN EN ISO 9241-210 (2010). Prozess zur Gestaltung gebrauchstauglicher interaktiver Systeme. ISO.

Dixon, Chris (2015). The Full-Stack Startup. Everything we've said about this trend, all in one place. a16z.com.

Doctorow, Cory (2014). The internet as a force for liberation, not enslavement. nextco.nf/2m6ZZxx. ᴵNEXT

Doctorow, Cory (2015). Information Doesn't Want to be Free: Laws for the Internet Age. McSweeney's Publishing. ᴵNEXT

Dolata, Ulrich (2013). The Transformative Capacity of New Technologies: A Theory of Sociotechnical Change. Routledge.

Dreifuss, Henry (1955). Designing for People. Simon & Schuster.

Dyson, George (2012). Turing's Cathedral. Pantheon. ᴵNEXT

Dyson, George (2012). Turing's Cathedral. nextco.nf/2m7olnO. ᴵNEXT

Ellis, Sean (2010). Find a Growth Hacker for Your Startup. startup-marketing.com.

Eriksson, Martin (2015). The History and Evolution of Product Management. mindtheproduct.com.

Eyal, Nir (2013). Hooked: How to Build Habit-Forming Products. Portfolio.

Ferriss, Timothy (2016). Tools of Titans: The Tactics, Routines, and Habits of Billionaires, Icons, and World-Class Performers. Vermilion. ᴺᴱˣᵀ

Fielding, Roy (2000). Architectural Styles and the Design of Network-based Software Architectures. University of California, Irvine.

Foreman, John W. (2013). Data Smart: Using Data Science to Transform Information into Insight. Wiley.

Foster, Richard (2012). Creative Destruction Whips through Corporate America. Innosight Executive Briefing Winter 2012.

Frahm, Klaus Peter et al. (2016). The Product Field Reference Guide.

Frost, Brad (2016). Atomic Design.

Geest, Yuri van (2015). Exponential Organizations – The New Normal. nextco.nf/2m74YOK. ᴺᴱˣᵀ

Geest, Yuri van et al. (2014). Exponential Organizations. Diversion Books. |NEXT

Gengnagel, Christoph et al. (2015). Rethink! Prototyping: Transdisciplinary Concepts of Prototyping. Springer International Publishing.

Gorbis, Marina (2013). The Nature of the Future: Dispatches from the Socialstructed World. Free Press. |NEXT

Gorbis, Marina (2013). The Nature of the Future: The Socialstructed World. nextco.nf/2m7hvBF. |NEXT

Griffin, Tren (2016). Two Powerful Mental Models: Network Effects and Critical Mass. Andreessen Horowitz, a16z.com.

Grove, Andrew (1997). Only the Paranoid Survive. HarperCollins Business.

Hagel, John (2015). The Power of Platforms. Deloitte University Press.

Hammersley, Ben (2013). Approaching the Future: 64 Things You Need to Know Now for Then. Soft Skull Press. |NEXT

Hansen, Jared (2015). How to Jumpstart User Testing: 16 Tools to Craft Better Products. Silver Fox Marketing Group.

Laurent Haug (2015). The Ongoing Reinvention of Shopping. nextco.nf/2m8DoAI. |NEXT

Hinssen, Peter (2015). The Network Always Wins. nextco.nf/2m8RgLa. |NEXT

Hinssen, Peter (2015). The Network Always Wins: How to Influence Customers, Stay Relevant, and Transform Your Organization to Move Faster than the Market. Mcgraw-Hill Education. |NEXT

J

Jarvis, Jeff (2009). The Great Restructuring. nextco.nf/2m9jeX9.

Jarvis, Jeff (2009). What Would Google Do? HarperBusiness. ᴺᴱˣᵀ

Jobs, Steve (2007). iPhone Introduction.

K

Kaushik, Avinash (2009). Web Analytics 2.0: The Art of Online Accountability and Science of Customer Centricity. Sybex.

Keen, Andrew (2011). Why Data Must Remain Neutral. nextco.nf/2m9jaXn. ᴺᴱˣᵀ

Keen, Andrew (2015). The Internet Is Not the Answer. Atlantic Monthly Press. ᴺᴱˣᵀ

Kelley, Tom (2001). The Art of Innovation: Lessons in Creativity from Ideo, America's Leading Design Firm. Crown Business.

Knapp, Jake et al. (2016). Sprint: How to Solve Big Problems and Test New Ideas in Just Five Days. Simon & Schuster.

Krishna, Golden (2015). The Best Interface Is No Interface: The simple path to brilliant technology. New Riders. ᴺᴱˣᵀ

Krishna, Golden (2016). The Best Interface Is No Interface. nextco.nf/2m998pc. ᴺᴱˣᵀ

Kumar, Vijay (2012). 101 Design Methods: A Structured Approach for Driving Innovation in Your Organization. Wiley.

Lacy, Sarah (2008). The Stories of Facebook, Youtube and Myspace: The People, the Hype and the Deals Behind the Giants of Web 2.0. Crimson Publishing. ˈNEXT

Lacy, Sarah (2011). Brilliant, Crazy, Cocky: How the Top 1% of Entrepreneurs Profit from Global Chaos. nextco.nf/2m9qthP. ˈNEXT

Laschke, Matthias et al. (2011). Things with attitude: Transformational Products. Conference Paper.

Leberecht, Tim (2015). The Business Romantic: Give Everything, Quantify Nothing, and Create Something Greater Than Yourself. HarperBusiness. ˈNEXT

Leberecht, Tim (2015). The Future of Business is Romantic. nextco.nf/2m9wIlx. ˈNEXT

Lockwood, Thomas (2009). Design Thinking: Integrating Innovation, Customer Experience, and Brand Value. Allworth Press.

Lucier, Charles E. et al. (1997). 10x Value: The Engine Powering Long-Term Shareholder Returns. strategy+business, Third Quarter 1997/Issue 8 (originally published by Booz & Company).

Mattin, David (2016). Trendwatching 2017. nextco.nf/2m9c9FP. ˈNEXT

Mattin, David et al. (2015). Trend-Driven Innovation: Beat Accelerating Customer Expectations. Wiley. ˈNEXT

Maurya, Ash (2012). Running Lean. O'Reilly Media.

McElroy, Kathryn (2017). Prototyping for Designers. Developing the Best Digital and Physical Products. O'Reilly Media.

Nahai, Nathalie (2012). Webs of Influence: The Secret Strategies That Make Us Click. Pearson.

Nahai, Nathalie (2016). The Psychology Behind Successful Products. nextco.nf/2m9tzlX.

Negroponte, Nicholas (1995). Being Digital. Alfred A. Knopf.

Norman, Donald A. (1988). The Design of Everyday Things. Basic Books.

Nudelman, Greg (2014). The $1 Prototype: Lean Mobile UX Design and Rapid Innovation for Material Design, iOS8, and RWD. DesignCaffeine Press.

Osterwalder, Alexander et al. (2010). Business Model Generation: A Handbook For Visionaries, Game Changers, And Challengers. Wiley.

Parker, Geoffrey G. et al. (2016). Platform Revolution. W. W. Norton & Company.

Peppers, Don (2013). Explaining Customer Centricity With a Diagram. LinkedIn Pulse.

Polaine, Andy (2013). Service Design: From Insight to Implementation. Rosenfeld Media.

Poppendieck, Mary et al. (2003). Lean Software Development: An Agile Toolkit. Addison-Wesley Professional.

Prendiville, Alison (2016). Connectivity through Service Design.
In: The Routledge Companion to Design Studies. Edited by Sparke, Penny
et al. Routledge.

Ries, Eric (2011). The Lean Startup: How Today's Entrepreneurs Use
Continuous Innovation to Create Radically Successful Businesses.
Crown Business.

Ritter, Frank E. et al. (2014). Foundations for Designing User-Centered
Systems. Springer London.

Scoble, Robert (2013). The Age of Context. nextco.nf/2m9rCpH. ^{NEXT}

Scoble, Robert et al. (2016). The Fourth Transformation: How Augmented
Reality & Artificial Intelligence Will Change Everything. Patrick Brewster Press. ^{NEXT}

Sehested, Claus et al. (2011). Lean Innovation: A Fast Path from Knowledge
to Value. Springer-Verlag.

Shapiro, Carl et al. (1999). Information Rules. A Strategic Guide to the
Network Economy. Harvard Business School Press.

Simon, Herbert A. (1962). The Architecture of Complexity. Proceedings of
the American Philosophical Society, Vol. 106, No. 6 (Dec. 12, 1962), pp. 467–482.

Simonds, Francesca (2016). Human Centred Design vs Design Thinking vs
Service Design vs UX …. What do they all mean? LinkedIn Pulse.

Skarin, Mattias (2015). Real-World Kanban: Do Less, Accomplish More with
Lean Thinking. Pragmatic Bookshelf.

Skok, Michael (2012). Must-read for founders: A VC explains how to build a killer value proposition. VentureBeat.

Smith, Adam (1776). An Inquiry into the Nature and Causes of the Wealth of Nations. W. Strahan and T. Cadell.

Smith, Benjamin (2015). A New Business Model for the Web? The Subscription Wars Are Here. Observer.com.

Solis, Brian (2015). X: The Experience When Business Meets Design. Wiley. ⁱNEXT

Solis, Brian (2016). The Future of Brand, Tech & Business is Experience. nextco.nf/2m9tazY. ⁱNEXT

Stellma, Andre et al. (2013). Learning Agile: Understanding Scrum, XP, Lean, and Kanban. O'Reilly Media.

Sterling, Bruce (2005). Shaping Things. The MIT Press. ⁱNEXT

Sterling, Bruce (2013). Fantasy Prototypes and Real Disruption. nextco.nf/2m9t23v. ⁱNEXT

Stickdorn, Marc et al. (2011). This is Service Design Thinking. Wiley.

Sutherland, Jeff (2014). Scrum: The Art of Doing Twice the Work in Half the Time. Crown Business.

Tariq, Ali Rushdan (2015). A Brief History of User Experience. blog.invisionapp.com.

Tetzeli, Rick (2016). Playing The Long Game Inside Tim Cook's Apple. in: Fast Company, September 2016.

Thiel, Peter et al. (2014). Zero to One: Notes on Startups, or How to Build the Future. Crown Business.

Thompson, Ben (2013). What Clayton Christensen Got Wrong. Stratechery.

Thompson, Ben (2015). Aggregation Theory. Stratechery.

Vargo, Stephen L. and Lusch, Robert F. (2004). Evolving to a New Dominant Logic for Marketing. Journal of Marketing 68 (January): pp. 1–17.

Vargo, Stephen L. and Lusch, Robert F. (2008). Service-Dominant Logic: Continuing the Evolution. Journal of the Academy of Marketing Science 36, pp. 1–10.

Vargo, Stephen L. and Lusch, Robert F. (2011). It's all B2B...and beyond: Toward a systems perspective of the market. Industrial Marketing Management 40, pp. 181–187.

Warfel, Todd Zaki (2009). Prototyping: A Practitioner's Guide. Rosenfeld Media.

Weinberger, David (2012). Too Big to Know. Basic Books. ⌐NEXT⌐

Weinberger, David (2012). Unsettling Knowledge. nextco.nf/2m9sUko. ⌐NEXT⌐

Wiggins, Adam (2011). The Twelve-Factor App. 12factor.net.

Wolfram, Stephen (2013). The Computational Knowledge Revolution. nextco.nf/2m9kgCp. ⌐NEXT⌐

Wolfram, Stephen (2016). Idea Makers: Personal Perspectives on the Lives & Ideas of Some Notable People. Wolfram Media. ᴺᴱˣᵀ

Wolpers, Stefan (2015). Lean User Testing: A Pragmatic Step-by-Step Guide to User Tests. Berlin Product People.

Speakers of the NEXT Conference are highlighted with ᴺᴱˣᵀ. As the leading conference for digital transformation in Germany, NEXT has been providing an inspiring way of understanding what will move consumers in the near future since 2006. The conference is embedded in the Hamburg Reeperbahn Festival with more than 40,000 participants and 800 concerts, events and conferences. nextconf.eu

1st edition, 2017

CC BY-NC-ND 4.0
Matthias Schrader, SinnerSchrader

Author

Matthias Schrader

Publisher

Next Factory Ottensen
SinnerSchrader Aktiengesellschaft
Völckersstr. 38
22765 Hamburg
Germany
+49 40 39 88 55 0
nextfactory@sinnerschrader.com

Translation

Tim Cole, Eric Doyle (editor)

Design/Creative Direction

Stellavie – Heidemann und Klein GbR

Print/Production

Kösel GmbH & Co. KG, Altusried

Material/Paper

Winter+Company Skivertex Matara
Salzer Design, 1,5-fach, 100 g/m²

Typefaces/Fonts

Noe Text, Maison, Maison Mono

ISBN

978-3-9818711-2-8
978-3-9818711-7-3 (E-Book)

goto 1